INDUSTRIAL MOTION CONTROL

INDUSTRIAL MOTION CONTROL

MOTOR SELECTION, DRIVES, CONTROLLER TUNING, APPLICATIONS

Hakan Gürocak

Washington State University
Vancouver, USA

Library of Congress Cataloging-in-Publication Data applied for.

A catalogue record for this book is available from the British Library.

ISBN: 9781118350812

Typeset in 10/12pt TimesLTStd by SPi Global, Chennai, India

Printed in the UK

Contents

Preface

Over the past couple of decades, the academic community has made significant advances in developing educational materials and laboratory exercises for fundamental mechatronics and controls education. Students learn mathematical control theory, board-level electronics, interfacing, and microprocessors supplemented with educational laboratory equipment. As new mechanical and electrical engineering graduates become practicing engineers, many are engaged in projects where knowledge of industrial motion control technology is an absolute must since industrial automation is designed primarily around specialized motion control hardware and software.

This book is an introduction to industrial motion control, which is a widely used technology found in every conceivable industry. It is the heart of just about any automated machinery and process. Industrial motion control applications use specialized equipment and require system design and integration where control is just one aspect. To design such systems, engineers need to be familiar with industrial motion control products; be able to bring together control theory, kinematics, dynamics, electronics, simulation, programming and machine design; apply interdisciplinary knowledge; and deal with practical application issues. Most of these topics are already covered in engineering courses in typical undergraduate curricula but in a compartmentalized nature, which makes it difficult to grasp the connections between them.

As I wrote this book, my goal was to bring together theory, industrial machine design examples, industrial motion control products and practical guidelines. The context of studying industrial motion control systems naturally brought separately taught topics together and often crossed disciplinary lines. The content came from my personal experience in developing and teaching mechatronics and automation courses, working with undergraduate students and from many discussions with engineers in the motion control industry. For example, even though many types of motors are available, I chose to concentrate on three-phase AC servo and induction motors based on input from the motion control industry. By no means this is a comprehensive book on any of the topics covered. It is not an in-depth examination of control theory, motor design, or power electronics. Rather, it is a balanced coverage of theory and practical concepts. Much of this material is available in manufacturer data sheets, manuals, product catalogs, fragments in various college courses, websites, trade magazines, and as know-how among practicing engineers. The book presents these pieces in a cohesive way to provide the fundamentals while supplementing them with solved examples based on practical applications.

The book starts with an introduction to the building blocks of a typical motion control system in Chapter 1. A block diagram is provided and the basic function of each building block is explained.

Chapter 2 examines how the motion profile is generated when an axis of a machine makes a move. After an overview of basic kinematics, two common motion profiles are explained. The chapter concludes with two approaches for multiaxis coordination.

As the mechanical design of each axis and the overall machine are significant factors in achieving the desired motion, Chapter 3 focuses on drive-train design. Concepts of inertia reflection, torque reflection, and inertia ratio are introduced. Five types of transmission mechanisms are explored in depth. Torque–speed curves of motors, gearboxes, and motor selection procedures for different types of motors and axes with transmission mechanisms are provided.

Electric motors are by far the most commonly used actuators in industrial motion control. Chapter 4 begins with fundamental concepts such as electrical cycle, mechanical cycle, poles, and three-phase windings. Construction and operational details of AC servo and induction motors are provided. Torque generation performance of AC servo motors with sinusoidal and six-step commutation is compared. The chapter concludes with mathematical and simulation models for both types of motors.

Motion control systems employ an assortment of sensors and control components along with the motion controller. Chapter 5 starts with the presentation of various types of optical encoders for position measurement, limit switches, proximity sensors, photoelectric sensors, and ultrasonic sensors. Sinking or sourcing designations for sensor compatibility to I/O cards are explained. Next, control devices including push buttons, selector switches, and indicator lights are presented. The chapter concludes with an overview of motor starters, contactors, overload relays, soft-starters, and a three-wire motor control circuit.

A drive is the link between the motor and the controller. It amplifies small command signals generated by the controller to high-power voltage and current levels necessary to operate a motor. Chapter 6 begins by presenting the building blocks of drive electronics. The popular pulse width modulation (PWM) control technique is explained. Then, basic closed-loop control structures implemented in the drive are introduced. Single-loop PID position control and cascaded velocity and position loops with feedforward control are explored in depth. Mathematical and simulation models of the controllers are provided. Control algorithms use gains that must be tuned so that the servo system for each axis can follow its commanded trajectory as closely as possible. The chapter concludes by providing tuning procedures for the control algorithms presented earlier and includes practical ways to address integrator saturation.

The book concludes with Chapter 7, which is about programming and motion control applications. Linear, circular, and contour move modes of a motion controller are explored. The chapter continues by introducing algorithms for basic programmable logic controller (PLC) functionality that are commonly used in motion controller programs. The chapter concludes by reviewing how a motion controller can control a non-Cartesian machine, such as a robot, by computing its forward and inverse kinematics in real-time.

One of the challenges in writing a book like this is the variety of motion controller hardware and software in the market and their proprietary nature. Each controller manufacturer has its own programming language and programming environment for their products. Since the programming details are very specific to each hardware, I attempted to provide algorithmic outlines rather than complete programs in a specific programming language or structure. The product manuals and manufacturer suggestions must be closely followed in adapting these algorithms to a specific choice of motion controller hardware. Another challenge was the digital control systems implemented in the controllers. These sampled data systems would be modeled using the z-transform (z-domain). However, almost all undergraduate engineering programs include only coverage of the continuous-time systems modeled using the Laplace transform (s-domain). Therefore, I chose to use the s-domain in presenting the control system models. This approach provides a good approximation since today's controllers have very fast sampling frequencies and the mechanical systems they control have relatively slow dynamics.

As I presented concepts, especially for the solved drive-train examples, I used data for industrial products from datasheets and catalogs. Today, most of these resources are available on the Internet as provided in the references of the chapters. Over time, manufacturers will change their products and these catalogs may not be available anymore. However, the theoretical coverage and the practical selection procedures should equally apply to the similar, newer future products.

This book is intended to be an introduction to the topic for undergraduate mechanical and electrical engineering students. Since many practicing engineers are involved in motion control systems, it should also be a resource for system design engineers, project managers, industrial engineers, manufacturing engineers, product managers, field engineers, mechanical engineers, electrical engineers, and programmers in the industry. For example, the tuning procedures in Chapter 4 were demonstrated using mathematical simulations. But if a real system with a motion controller is available, then these procedures can simply be used to tune the controller of the real system without the need for any of the simulations. Similarly, Chapter 7 provides algorithms for common types of motion control applications such as winding. These algorithms can be a starting point to develop the control programs in the programming language of the real system. As stated earlier, the product manuals and manufacturer suggestions must be closely followed in adapting these algorithms to a specific choice of real motion controller hardware.

I am indebted to many people who helped me through the journey of writing this book. Many thanks to Mr. Ken Brown, President of Applied Motion Systems, Inc., Mr. Ed Diehl, President of Concept Systems, Inc., for our discussions about motion control systems, for the valuable suggestions and materials they provided. Special thanks goes to Mr. Dimitri Dimitri, President of Delta Tau Data Systems, Inc., and Mr. Curtis Wilson, Vice President of Engineering and Research at Delta Tau Data Systems, Inc. for their contributions to our laboratory, to this book and for their in-depth technical guidance. I would also like to acknowledge the contributions by Mr. Dean Ehnes, Mechanical Engineer, Mr. John Tollefson, Electrical Engineer, and Mr. Brian Hutton, General Manager of Columbia/Okura, LLC through insightful discussions about material handling systems, motors, and for ideas for example problems. My colleague Dr. Xiaodong Liang from electrical engineering provided a detailed review of the chapters on motors and electric drives, which was much appreciated. I received tremendous help from Mr. Ben Spence, my former student from a long time ago and now Systems Engineer at Applied Motion Systems, Inc. I learned a lot from him as we spent countless hours working together on laboratory machines and discussing many technical aspects of motion control. Also, I am grateful to Mr. Matthew Bailie, Mechanical Engineer at Applied Motion Systems, Inc. for his guidance about the motor sizing procedures and for ideas for some of the example problems. Many thanks to the companies that provided the product photographs used in the book. I would like to thank Mr. Paul Petralia, Senior Editor at John Wiley & Sons and the Wiley staffs Sandra Grayson, Clive Lawson and Siva Raman Krishnamoorty, for their guidance throughout the process of developing this book. I had the privilege of working with many outstanding students who learned this material as I tried it out in courses I taught in the last few years. I really appreciated their valuable feedback, suggestions, patience, and enthusiasm.

As I discovered through this process, writing a textbook is a major undertaking. I am very grateful to my wife for her endless patience, support, and encouragement while I was writing this book for the last 3 years.

Hakan Gürocak
Vancouver, Washington, USA

1

Introduction

Motion control is widely used in all types of industries including packaging, assembly, textile, paper, printing, food processing, wood products, machinery, electronics, and semiconductor manufacturing. It is the heart of just about any automated machinery and process. Motion control involves controlling mechanical movements of a load. For example, in case of an inkjet printer, the load is the ink cartridge that has to be moved back-and-forth across the paper with high speed and precision. On the other hand, in a paper-converting machine, the load can be the large parent roll of paper that is loaded into the machine for processing. In this case, the load rotates as the paper is unwound from the roll and rewound into smaller processed rolls such as embossed paper towels.

Each motor moves a segment of the mechanical components of the machine. The segment of the machine along with the motor that moves it is called an *axis*. Considering an inkjet printer as an example, the mechanical components involved in the sliding motion of the print cartridge and the motor driving them collectively make up an axis of the machine. Another axis of the printer consists of all mechanical components and the motor that feed paper into the printer. In case of the paper-converting machine, the mandrel that holds the roll of paper, the pulleys, and belts that connect it to the motor and the motor make up an axis.

A typical motion control system manages position, velocity, torque, and acceleration of an axis. Often the machine consists of multiple axes whose position and/or velocity must be controlled in a synchronized fashion. For example, the X-axis and Y-axis of the table of a CNC machine need to be controlled in a coordinated way so that the machine can cut a round corner in the work piece. The ability to precisely control and coordinate complex motions of multiple axes enables design of industrial machines such as those in Figure 1.1.

Prior to the programmable motion controllers, coordination was achieved through mechanical means [14]. A central line shaft was connected to a large electric motor or an engine that ran at constant speed. This motion source was then used to drive all the axes of the machine by coupling them to the line shaft through pulleys, belts, gears, cams, and linkages (Figure 1.2). Clutches and brakes were used to start or stop the individual axes. The gear ratios between the line shaft and the individual axes determined the speed of each axis. Drive-trains, which were often long shafts, transferred the coordinated motion to the appropriate part of the

Industrial Motion Control: Motor Selection, Drives, Controller Tuning, Applications, First Edition. Hakan Gürocak.
© 2016 John Wiley & Sons, Ltd. Published 2016 by John Wiley & Sons, Ltd.

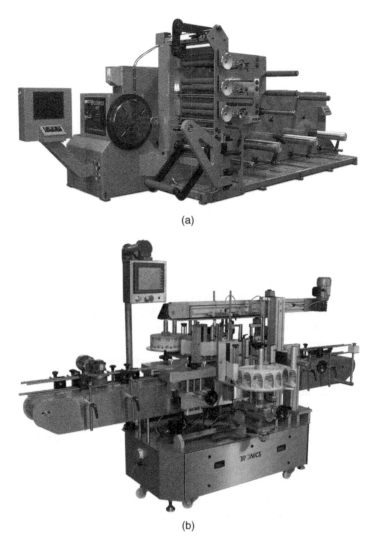

(a)

(b)

Figure 1.1 Complex machines with multiple axes are made possible with the ability of the controller to precisely coordinate motion of all axes. (a) Foil and wire winding machine (Reproduced by permission of Broomfield, Inc. [3]). (b) Pressure sensitive labeling machine (Reproduced by permission of Tronics America, Inc. [19])

machine. Complex machinery required sophisticated mechanical designs. Backlash, wear, and deflections in the long shafts were problematic. The biggest challenge was when a change in the product had to be introduced into the production system. It required physically changing the gear reducers, which was costly and very time consuming. Also, realigning the machine for accurate timing was difficult after drive-train changes.

As computers became main-stream equipment through the inexpensive availability of electronics, microprocessors, and digital signal processors, coordination of motion in multiaxes

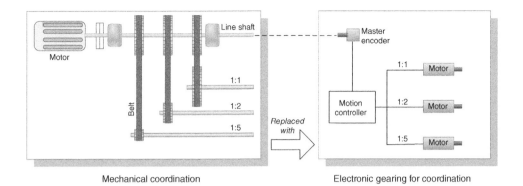

Figure 1.2 Multiaxis coordination

machines began to shift into a computer-controlled paradigm. In a modern multiaxis machine, each axis has its own motor and electric drive. Coordination between the axes is now achieved through electronic gearing in software. The drive-trains with long shafts are replaced by short and much more rigid shafts and couplings between the motor and the mechanism it drives. The motion controller interprets a program and generates position commands to the drives of the axes. These motion profiles are updated in real-time as the drives commutate the motors and close the control loops. In today's technology it is typical for an ordinary motion controller to coordinate up to eight axes at a time. Controllers with 60+ axis capabilities are available.

1.1   Components of a Motion Control System

The complex, high-speed, high-precision control required for the multiaxis coordinated motion is implemented using a specialized computer called **motion controller**. As shown in Figure 1.3, a complete motion control system consists of:

1. Human–machine interface (HMI),
2. Motion controller
3. Drives
4. Actuators
5. Transmission mechanisms
6. Feedback.

1.1.1   Human–Machine Interface

The HMI is used to communicate with the motion controller. The HMI may serve two main functions: (i) Operating the machine controlled by the motion controller, and (ii) Programming the motion controller.

Control panels as shown in Figure 1.4a with pilot lights, push buttons, indicators, digital readouts, and analog gauges are common hardware-based HMIs to serve the purpose of operating a machine. Chapter 5 discusses operator interface devices such as pilot lights, push buttons,

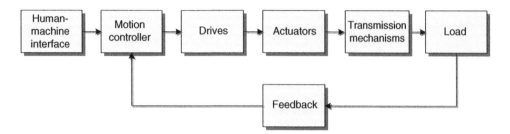

Figure 1.3 Components of a motion control system

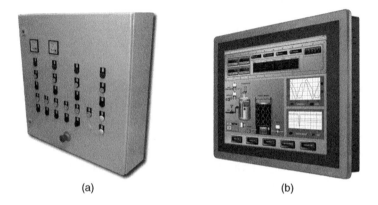

        (a)   (b)

Figure 1.4 Human–machine interfaces (operator panels) are used to operate the machine. (a) Hardware-based control panel [7]. (b) Software-based control panel with touch screen (Reproduced by permission of American Industrial Systems, Inc. [2])

and selector switches. Control panels can also be software-based as shown in Figure 1.4b. These panels can have touch screens and embedded computers that run a graphical user interface developed using software [10, 13, 15]. The advantage of this type of panel is the ease of reconfiguration of the HMI as new features may be added to the machine in the future.

A computer is interfaced to the motion controller for programming purposes. Custom software provided by the manufacturer of the controller is used to write, edit, download, and test the machine control programs. The software also includes features to test motors, monitor input/output (I/O) signals, and tune controller gains. Chapter 7 presents programming approaches for motion, machine I/O management, and multiaxis coordination.

1.1.2 Motion Controller

The motion controller is the "brains" of the system. It generates motion profiles for all axes, monitors I/O, and closes feedback loops. As presented in Chapter 2, the controller generates the motion profile for an axis based on the desired motion parameters defined by the user or the programmer. While the machine is running, it receives feedback from each axis motor. If there is a difference (following error) between the generated profile and the actual position or

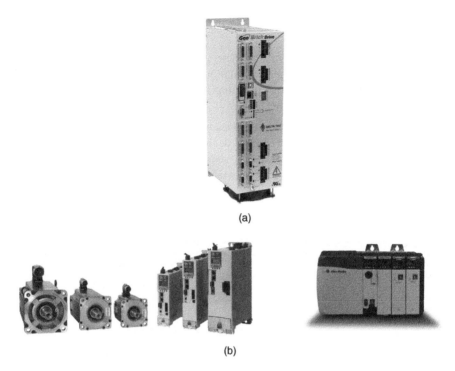

(a)

(b)

Figure 1.5 Motion controllers can have the various form factors. (a) Integrated form factor (Reproduced by permission of Delta Tau Data Systems, Inc. [5]). (b) Modular form factor. (Left) motors and drives [16], (Right) controller [17] (Courtesy of *Rockwell Automation, Inc.*)

velocity of an axis, the controller generates correction commands, which are sent to the drive for that axis. Chapter 6 discusses various control algorithms used to act on the following error to generate command signals to eliminate the error. As discussed in Chapter 7, the controller can also generate and manage complex motion profiles including electronic camming, linear interpolation, circular interpolation, contouring, and master–slave coordination.

Motion controllers are available in different form factors (Figure 1.5). The integrated form factor incorporates the computer, the drive electronics for the axes, and the machine I/O into a single unit. This unit is called motion controller or drive. In a modular system, the computer, the drives, and the machine I/O are separate units connected to each other via some type of communication link. In this case, just the computer is called the motion controller.

A complete motion controller consists of the following:

1. Computer
   - Interpretation of user programs
   - Trajectory generation
   - Closing the servo loops
   - Command generation for the drives (amplifiers)
   - Monitoring axis limits, safety interlocks
   - Handling interrupts and errors such as excessive following (position) error.

2. I/O for each axis
   - Motor power output
   - Servo I/O for command output to amplifiers
   - Input terminals for feedback signals from motor or other external sensors
   - Axis limits, homing signals, and registration.
3. Machine I/O
   - Digital input terminals for various sensors such as operator buttons and proximity sensors
   - Digital output terminals to drive external devices (usually through relays)
   - Analog inputs (often optional) for analog sensors such as pressure, force
   - Analog outputs (often optional) to drive analog devices.
4. Communication
   - Network communications with other peripheral devices, the host computer, and/or supervisory system of the plant using protocols such as DeviceNet®, Profibus®, ControlNet®, EtherNet/IP®, or EtherCAT®
   - USB or serial port communications, and
   - HMI communications.

1.1.3 Drives

The command signals generated by the controller are small signals. The drive (Figure 1.6) amplifies these signals to high-power voltage and current levels necessary to operate a motor. Therefore, the drive is also called an amplifier. The drive closes the current loop of the servo system as discussed in Chapter 6. Therefore, it must be selected to match the type of motor

(a)   (b)

Figure 1.6 Drives are used to provide high voltage and current levels necessary to operate motors. (a) Digital servo drive (Reproduced by permission of ADVANCED Motion Controls®) [1]. (b) AC drive (Courtesy of *Rockwell Automation, Inc.*) [18]

to be driven. In recent trends, the line between a drive and a controller continues to blur as the drives perform many of the complex functions of a controller. They are expected to handle motor feedback and not only close the current but also the velocity and position loops.

1.1.4   Actuators

An actuator is a device that provides the energy to move a load. Motion control systems can be built using hydraulic, pneumatic, or electromechanical (motor) technologies. This book presents three-phase AC servo and induction motors (Figure 1.7). Underlying concepts of electromechanical operation of these motors along with mathematical models are presented in Chapter 6. Specific control algorithms implemented in the drive to control each type of motor are explored in Chapter 6. When a machine for a motion control application is designed, motors must be carefully selected for proper operation of the machine. Chapter 3 presents torque–speed curves for each of these motors and design procedures for proper motor sizing.

1.1.5   Transmission Mechanisms

A transmission mechanism is used to connect the load to the motor of an axis. It helps meet the motion profile requirements. Chapter 3 presents gearboxes (Figure 1.8), lead/ball-screw drives, linear belt drives, pulley-and-belt drives, and conveyors. When a load is coupled to a motor through a transmission mechanism, the load inertia and torque are reflected through the mechanism to the motor. Chapter 3 provides extensive discussion on the mathematical models for this. Motor, gearbox, and transmission selection procedures are also provided.

1.1.6   Feedback

Feedback devices are used to measure the position or speed of the load. Also, the drive and the controller use feedback to determine how much current needs to be applied to each phase of the motor as explained in detail in Chapters 4 and 6. Most common feedback devices are resolvers, tachometers, and encoders. Chapter 5 explains encoders, which can be rotary or

(a)   (b)

Figure 1.7   AC servo and induction motors are used in motion control applications as actuators. (a) AC servo motors (Reproduced by permission of Emerson Industrial Automation [8]). (b) AC induction motor (Reproduced by permission of Marathon™ Motors, A Regal Brand) [12]

(a)   (b)

Figure 1.8 Gearboxes are used in motion control applications to help achieve speed and torque requirements. (a) In line gearhead for servo motors (Reproduced by permission of DieQua Corp. [6]). (b) Right-angle worm gear reducer for AC induction motors (Reproduced by permission of Cone Drive Operations, Inc. [4])

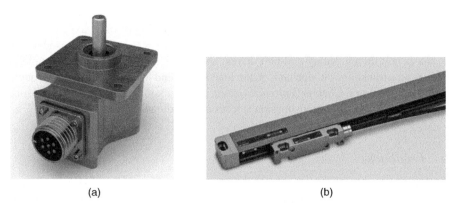

(a)   (b)

Figure 1.9 Encoders are used in motion control applications as feedback devices. (a) Rotary encoder (Reproduced by permission of US Digital Corp. [20]). (b) Linear encoder (Reproduced by permission of Heidenhain Corp. [9])

linear as shown in Figure 1.9. In addition, encoders can be incremental or absolute. Selection of the feedback device depends on the desired accuracy, cost, and environmental conditions of the machine.

A different type of feedback is provided to the controller from detection sensors such as proximity switches, limit switches, or photoelectric sensors. These devices detect presence or absence of an object. For example, a photoelectric sensor such as the one in Figure 1.10 may detect arrival of a product on a conveyor and signal the motion controller to start running the conveyor.

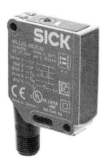

Figure 1.10 Photoelectric sensors are used for detection of presence of an object [11]

References

[1] Advanced Motion Controls (2014). DPCANIE-C060A400 Digital Servo Drive. http://www.a-m-c.com/index .html (accessed 21 November 2014).

[2] American Industrial Systems, Inc. (AIS) (2013). Ip17id7t-m1-5rt operator interface computer. http://www .aispro.com/products/ip17id7t-m1-5rt (accessed 3 November 2014).

[3] Broomfield (2014). 800 HV/LV Wire and Foil Winder with Optional Touch Screen Controller. http://www .broomfieldusa.com/foil/hvlv/ (accessed 31 October 2014).

[4] Cone Drive Operations, Inc. (2013) Cone Drive® Model RG Servo Drive. http://conedrive.com/Products /Motion-Control-Solutions/model-rg-gearheads.php (accessed 28 October 2014).

[5] Delta Tau Data Systems, Inc. Geo Brick Drive (2013). http://www.deltatau.com/DT_IndexPage/index.aspx (accessed 22 September 2013).

[6] DieQua Corp. (2014). Planetdrive – Economical Servo Gearheads. http://www.diequa.com/products/planetroll /planetroll.html (accessed 20 November 2014).

[7] Elmschrat (2010). File:00-bma-automation-operator-panel-with-pushbuttons.jpg. http://commons.wikimedia .org/wiki/File:00-bma-automation-operator-panel-with-pushbuttons.JPG (accessed 18 November 2014).

[8] Emerson Industrial Automation (2014). Servo Motors Product Data, Unimotor HD. http://www .emersonindustrial.com/en-EN/documentcenter/ControlTechniques/Brochures/CTA/BRO_SRVMTR_1107.pdf (accessed 14 November 2014).

[9] Heidenhain, Corp. (2014). Incremental Linear Encoder MSA 770. http://www.heidenhain.us (accessed 28 October 2014).

[10] Invensys, Inc. (2014). Wonderware InTouch HMI. http://global.wonderware.com/DK/Pages/default.aspx (accessed 4 November 2014).

[11] Lucasbosch (2014). SICK WL12G-3B2531 photoelectric reflex switch angled upright.png. http://commons .wikimedia.org/wiki/File:SICK_WL12G-3B2531_Photoelectric_reflex_switch_angled_upright.png (accessed 7 November 2014).

[12] Marathon Motors (2014). Three Phase Globetrotter® NEMA Premium® Efficiency Totally Enclosed AC Motor. http://www.marathonelectric.com/motors/index.jsp (accessed 5 November 2014).

[13] Progea USA, LLC (2014). Movicon™11. http://www.progea.us (accessed 4 November 2014).

[14] John Rathkey (2013). Multi-axis Synchronization. http://www.parkermotion.com/whitepages/Multi-axis.pdf (accessed 15 January 2013).

[15] Rockwell Automation (2014). FactoryTalk View™. http://www.rockwellautomation.com (accessed 17 November 2014).

[16] Rockwell Automation, Inc. (2014). Allen-Bradley® Kinetix 5500 servo drive, Kinetix VPL servo motor and single-cable technology. http://ab.rockwellautomation.com/Motion-Control/Servo-Drives/Kinetix-Integrated -Motion-on-EIP/Kinetix-5500-Servo-Drive (accessed 17 November 2014).

[17] Rockwell Automation, Inc. (2014). ControlLogix® Control System. http://ab.rockwellautomation.com /Programmable-Controllers/ControlLogix (accessed 17 November 2014).

[18] Rockwell Automation, Inc. (2014). PowerFlex® 525 AC Drives. http://ab.rockwellautomation.com/drives /powerflex-525 (accessed 17 November 2014).

[19] Tronics America, Inc. (2014). Series 3 premier pressure sensitive labeling system. http://www.tronicsamerica .com/index.htm (accessed 25 October 2014).

[20] US Digital Corp. (2014). HD25 Industrial Rugged Metal Optical Encoder. http://www.usdigital.com (accessed 17 November 2014).

2

Motion Profile

A moving object follows a trajectory. In an automatic machine, motion may involve a single axis moving along a straight line. In more complex cases, such as moving the cutting tool of a CNC milling machine along a circular path, coordinated motion of multiple axes is required.

When an axis of the machine needs to move from point "A" to "B", a trajectory connecting these points needs to be generated. In motion control, the trajectory is also called *motion profile*. The motion profile should lead to smooth acceleration of the axis from point "A" to a constant operational speed. After moving at this speed for a while, the axis needs to smoothly decelerate to come to a stop at point "B".

Motion controller generates the motion profile at regular intervals to create velocity and position commands for the servo control system of each motor. Each servo control system then regulates its motor signals to move its axis along the desired profile.

This chapter begins with an overview of basic kinematics. Then, the two most common motion profiles, namely trapezoidal and S-curve velocity profiles, are presented. The chapter concludes with a discussion of the slew and interpolated motion approaches for multiaxis coordination.

2.1 Kinematics: Basic Concepts

Kinematics is the study of motion without considering the forces causing that motion. It governs the relationships between time, position, velocity, and acceleration. Studying the kinematics of a machine is needed for motion profile calculations but also in choosing the right motors for the axes during the design of the machine.

As an axis is moving from point "A" to "B", its position, $s(t)$, along the trajectory is a function of time. Velocity, $v(t)$, is the change of position $s(t)$ in a given time interval. It is defined as

$$v(t) = \frac{ds}{dt}$$

Similarly, acceleration, $a(t)$, is the change of velocity $v(t)$ in a given time interval:

$$a(t) = \frac{dv}{dt} \tag{2.1}$$

Industrial Motion Control: Motor Selection, Drives, Controller Tuning, Applications, First Edition. Hakan Gürocak.
© 2016 John Wiley & Sons, Ltd. Published 2016 by John Wiley & Sons, Ltd.

These equations can also be written in the integration form as

$$s = \int v(t)\, t \tag{2.2}$$

$$v = \int a(t)\, t \tag{2.3}$$

Recall that integration of a function is the summation of infinitesimal elements under the curve of the function. Therefore, it is equal to the area under the curve. Then, from Equation (2.2), position at time t is equal to the area under the velocity curve up to time t as shown in Figure 2.1.

Slope of a curve at a point can be found using differentiation. Then, from Equation (2.1) acceleration is the slope of the velocity curve as shown in Figure 2.1.

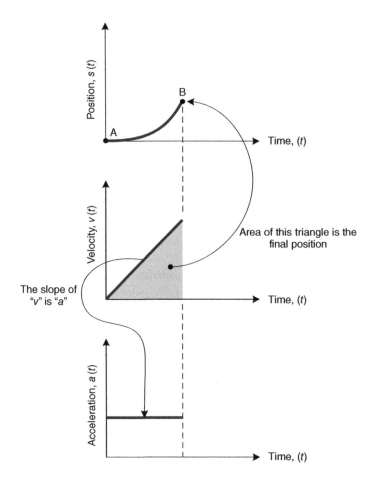

Figure 2.1   Basic relationships between position, velocity, and acceleration

> ### Geometric Rules for Motion Profile
>
> 1. Position at time t is equal to the area under the velocity curve up to time t
> 2. Acceleration is the slope of the velocity curve

For a more general case, the following can be derived from Equations (2.2) and (2.3):

$$v = v_0 + a(t - t_0)$$

$$s = s_0 + v_0(t - t_0) + \frac{1}{2}a(t - t_0)^2 \tag{2.4}$$

where t_0 is the initial time, v_0 is the initial velocity, and s_0 is the initial position. The acceleration "a" is constant.

■ EXAMPLE 2.1.1

Given the velocity profile in Figure 2.2, find the position and acceleration at $t = 5$ s.

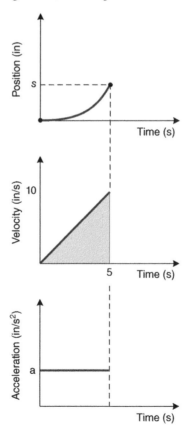

Figure 2.2   Velocity profile for acceleration

Solution

Slope of the velocity profile is the acceleration. Therefore,

$$a = \frac{10}{5}$$

$$= 2\, in/s^2$$

The triangular area under the velocity curve up to $t = 5$ s is the position reached at $t = 5$ s. Hence,

$$s = \frac{1}{2}(10 \cdot 5)$$

$$= 25\, in/s$$

■ **EXAMPLE 2.1.2**

An axis is traveling at a speed of 10 in/s. At $t = 5$ s it starts to slow down as given by the velocity profile in Figure 2.3. What is the axis position when it stops? Assume that the axis starts decelerating at 25 in.

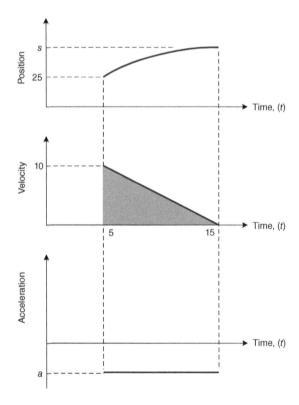

Figure 2.3   Velocity profile for deceleration

Solution

Slope of the velocity profile is the acceleration. In this case, the slope is negative since the axis is decelerating. Therefore,

$$a = -\frac{10}{10}$$
$$= -1 \, \text{in/s}^2$$

The triangular area under the velocity curve is the position reached at $t = 15$ s. Hence,

$$\Delta s = \frac{1}{2} \cdot 10(15 - 5)$$
$$= 50 \, \text{in}$$

The axis travels 50 in while it is slowing down. Since it started its deceleration at 25 in, the axis will be at $s = 75$ in when it stops.

2.2   Common Motion Profiles

There are two commonly used motion profiles:

- Trapezoidal velocity profile
- S-curve velocity profile.

The trapezoidal velocity profile is popular due to its simplicity. The S-curve velocity profile leads to smoother motion.

2.2.1   Trapezoidal Velocity Profile

Figure 2.4 shows the trapezoidal velocity profile, the resulting acceleration, and position profiles. It also shows the *jerk* (also called *jolt*), which is the derivative of acceleration. As it can be seen, the jerk is infinite at four points in this motion profile due to the discontinuities in the acceleration at corners of the velocity profiles. There are three distinct phases in the motion: (1) acceleration, (2) constant velocity (zero acceleration), and (3) deceleration.

To move an axis of the machine, usually the following desired motion parameters are known:

- Move velocity v_m
- Acceleration a
- Distance s to be traveled by the axis.

The desired motion profile can be programmed into the motion controller by first specifying the move velocity and move time for the motion. Then, the program commands the axis to move through distance s.

2.2.1.1 Geometric Approach

To calculate the move time t_m, we can apply the geometric rules to Figure 2.4 starting with the slope of the velocity curve $a = v_m/t_a$:

$$t_a = t_d = \frac{v_m}{a} \tag{2.5}$$

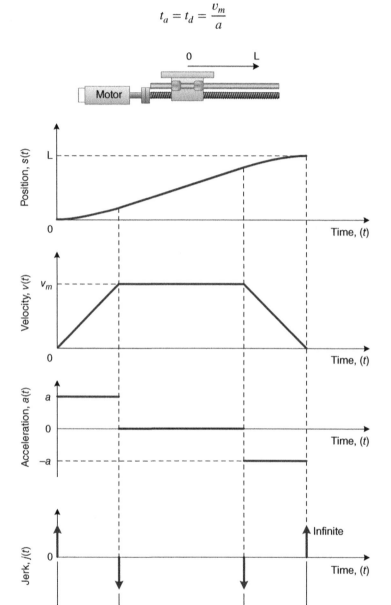

Figure 2.4 Trapezoidal velocity profile and associated position, acceleration, and jerk profiles to move an axis from 0 to position L

The acceleration and deceleration times do not have to be equal but this is often the case. The total motion time is as follows:

$$t_{\text{total}} = t_a + t_m + t_d \tag{2.6}$$

Using the geometric rules, the total distance traveled by the axis can be found by adding the areas of the two triangles and the rectangle under the velocity curve ($t_a = t_d$):

$$L = \frac{t_a v_m}{2} + t_m v_m + \frac{t_d v_m}{2}$$
$$= v_m(t_a + t_m)$$

The move time can then be found as

$$t_m = \frac{L}{v_m} - t_a \tag{2.7}$$

2.2.1.2 Analytical Approach

The motion controller can compute the axis position at any instant using Equation (2.4). Since the motion consists of three phases, this equation must be computed using the correct boundary values (t_0, v_0, s_0) in each segment of the motion. The move time must first be calculated using Equations (2.5) and (2.7).

ANALYTICAL APPROACH FOR TRAPEZOIDAL VELOCITY PROFILE

For $0 \leq t \leq t_a$

$t_0 = 0$, $v_0 = 0$, $s_0 = 0$

$$s(t) = \frac{1}{2}at^2 \tag{2.8}$$

For $t_a < t \leq (t_a + t_m)$

$t_0 = t_a$, $v_0 = v_m$, $s_0 = s(t_a)$, $a = 0$

$$s(t) = s(t_a) + v_m(t - t_a) \tag{2.9}$$

where $s(t_a)$ is the position found from Equation (2.8) at t_a.

For $(t_a + t_m) < t \leq t_{\text{total}}$

$t_0 = (t_a + t_m)$, $v_0 = v_m$, $s_0 = s(t_a + t_m)$

Continued

ANALYTICAL APPROACH FOR TRAPEZOIDAL VELOCITY PROFILE
(CONTINUED)

$$s(t) = s(t_a + t_m) + v_m[t - (t_a + t_m)]$$
$$-\frac{1}{2}a[t - (t_a + t_m)]^2$$

where $s(t_a + t_m)$ is the position found from Equation (2.9) at
$t = (t_a + t_m)$. Note that the acceleration is negative.

■ EXAMPLE 2.2.1

The X-axis of a gantry robot is to move for 10 in. The maximum acceleration allowed for this
axis is 1 in/s². If the axis needs to move at a desired maximum velocity of 2 in/s, how long
will it take to complete this motion?

Solution
From Equation (2.5), we can find the acceleration time as

$$t_a = t_d = \frac{v_m}{a} = \frac{2}{1} = 2\,\text{s}$$

From Equation (2.7)

$$t_m = \frac{L}{v_m} - t_a = \frac{10}{2} - 2 = 3\,\text{s}$$

Then, the total motion time from Equation (2.6) is

$$t_{total} = t_a + t_m + t_d = 2 + 3 + 2 = 7\,\text{s}$$

Figure 2.5 shows an outline of the velocity profile.

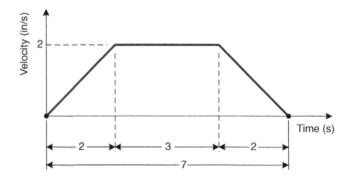

Figure 2.5   Trapezoidal velocity profile for the X-axis

■ EXAMPLE 2.2.2

Given the velocity profile in Figure 2.6, calculate s_A, s_B, s_C using the geometric rules for motion profile given in Section 2.1.

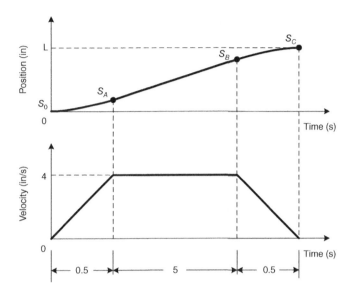

Figure 2.6   Trapezoidal velocity for Example 2.2.2

Solution

s_A can be found from the area of the triangle under the velocity curve as

$$s_A = \frac{1}{2} \cdot 4 \cdot 0.5$$
$$= 1 \text{ in}$$

s_B is the total area under the velocity curve up to point B. It is equal to the area of the left triangle plus the area of the rectangle.

$$s_B = s_A + 4 \cdot 5$$
$$s_B = \frac{1}{2} \cdot 4 \cdot 0.5 + 4 \cdot 5$$
$$= 21 \text{ in}$$

s_C is the total area under the velocity curve up to point C. It is equal to the area of the two triangles plus the area of the rectangle.

$$s_B = 2 \left(\frac{1}{2} \cdot 4 \cdot 0.5 \right) + 4 \cdot 5$$
$$= 22 \text{ in}$$

■ **EXAMPLE 2.2.3**

Given the same velocity profile in Example 2.2.2 calculate s_A, s_B, s_C using the analytical method described in Section 2.2.1.2.

Solution
From the velocity profile we can calculate the acceleration as $a = 4/0.5 = 8\,\text{in/s}^2$.

Acceleration phase $(0 \leq t \leq 0.5)$
   In this segment of the motion, $t_0 = 0, s_0 = 0, v_0 = 0, a = 8$. Therefore,

$$s(t) = \frac{1}{2}a(t - t_0)^2$$

$$s_A = \frac{1}{2} \cdot 8 \cdot 0.5^2$$

$$= 1\,\text{in}$$

Constant velocity phase $(0.5 < t \leq 5.5)$
   In this segment of the motion, $t_0 = 0.5, s_0 = s_A, v_0 = v_A, a = 0$. Therefore,

$$s(t) = s_A + v_A(t - t_0) + \frac{1}{2}a(t - t_0)^2$$

$$= 1 + 4(t - 0.5) + \frac{1}{2}0(t - 0.5)^2$$

$$s_B = 1 + 4(5.5 - 0.5)$$

$$= 21\,\text{in}$$

Deceleration phase $(5.5 < t \leq 6)$
   In this segment of the motion, $t_0 = 5.5, s_0 = s_B, v_0 = v_B, a = -8$. Therefore,

$$s(t) = s_B + v_B(t - t_0) + \frac{1}{2}a(t - t_0)^2$$

$$= 21 + 4(t - 5.5) - \frac{1}{2}8(t - 5.5)^2$$

$$s_C = 21 + 4(6 - 5.5) - \frac{1}{2}8(6 - 5.5)^2$$

$$= 22\,\text{in}$$

■ **EXAMPLE 2.2.4**

Given the symmetric triangular velocity profile in Figure 2.7, calculate the maximum velocity and acceleration.

Solution
Using the geometry (area under the velocity curve), we can calculate the total distance traveled as

$$s_B = \frac{1}{2}v_{\max}\frac{t}{2} + \frac{1}{2}v_{\max}\frac{t}{2}$$

$$= \frac{1}{2}v_{\max}t$$

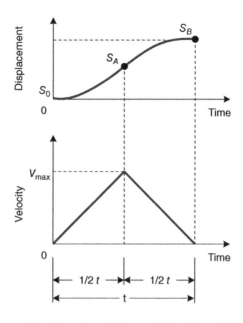

Figure 2.7   Triangular velocity profile for Example 2.2.4

Then,

$$v_{max} = \frac{2s_B}{t}$$

and

$$a = \frac{2v_{max}}{t}$$

2.2.2   S-curve Velocity Profile

The trapezoidal velocity profile is simple to compute but it has a major drawback. The sharp corners of the trapezoid cause discontinuities in the acceleration which then lead to infinite (or in practice large) jolts in the system.

To smooth out the acceleration, the sharp corners are rounded to create the so-called *S-curve velocity profile* shown in Figure 2.8. The rounded corners are made up of second-order polynomials. This reshaping of the velocity profile smooths the transitions between positive, zero, and negative acceleration phases. Note that unlike the trapezoidal velocity profile, the acceleration is not constant. Also, unlike the trapezoidal velocity profile, the jerk is not infinite. As long as the jerk is finite, there will be no sudden shock loads to disrupt smooth cycle operation [2]. Instantaneous changes in force/torque and current demands on the motor are eliminated. Additionally, high-frequency oscillatory motions are reduced. The motor's longevity and the system's accuracy are increased through the use of S-curve motion profile.

As shown in Figure 2.8, there are seven distinct phases represented in this velocity profile [4]. The four curved segments of the profile are implemented using quadratic equations. the remaining three segments are straight lines with positive, zero, and negative slope. If the linear segments are eliminated, we will end up with the *pure S-curve* velocity profile shown in

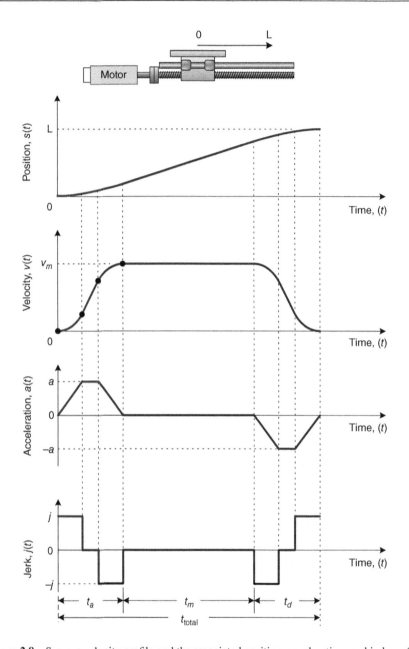

Figure 2.8 S-curve velocity profile and the associated position, acceleration, and jerk profiles

Figure 2.9. The pure S-curve is made up of two pieces of quadratic curves designated as "A" and "B". Each piece can be expressed by the following equation [3]:

$$v(t) = C_1 t^2 + C_2 t + C_3 \tag{2.10}$$

where C_1, C_2, and C_3 are coefficients to be determined using boundary conditions.

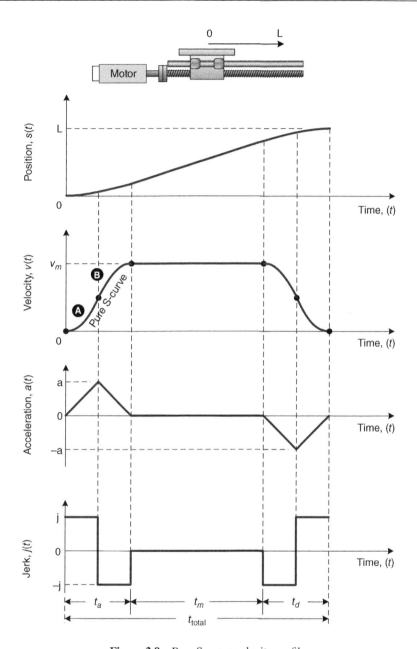

Figure 2.9 Pure S-curve velocity profile

Curve A: We can define the following boundary conditions for the "A" segment:

$$v(0) = 0 \tag{2.11}$$

$$a(0) = \frac{dv}{dt} = 0 \tag{2.12}$$

$$v\left(\frac{t_a}{2}\right) = \frac{v_m}{2} \tag{2.13}$$

$$a\left(\frac{t_a}{2}\right) = a \tag{2.14}$$

Using Equations (2.10) and (2.11), $C_3 = 0$ can be found. Similarly, after differentiating Equation (2.10) and using (2.12), we can find $C_2 = 0$. Then, the equation for curve A is found as $v(t) = C_1 t^2$. Evaluating this using Equation (2.13) gives the following:

$$v_m = \frac{1}{2} C_1 t_a^2$$

or,

$$C_1 = \frac{2v_m}{t_a^2} \tag{2.15}$$

Also, the maximum acceleration can be found by differentiating the equation for curve A and using (2.14) as

$$a = C_1 t_a \tag{2.16}$$

Solving for t_a from (2.16) and substituting into (2.15) gives

$$C_1 = \frac{a^2}{2v_m} \tag{2.17}$$

Hence, the equation for curve A in Figure 2.9 for $0 \le t \le \frac{t_a}{2}$ is

$$v(t) = \frac{a^2}{2v_m} t^2 \tag{2.18}$$

Distance traveled during the velocity curve A is the integral of Equation (2.18), which is

$$s_A(t) = \frac{a^2 t^3}{6v_m} \tag{2.19}$$

where it was assumed that the motion started from zero initial position. The acceleration time can be found from solving for C_1 from (2.16) and substituting into (2.15) as

$$t_a = \frac{2v_m}{a}$$

Curve B: The equation for the "B" segment is valid for $\frac{t_a}{2} < t \le t_a$. If we take the speed v_A at the end of segment "A" and add the current speed $v(t)$ at time t to it, we should ultimately get to v_m when $t = t_a$. Using this approach, the following equation can be written:

$$v_m = v_A + v(t)$$
$$v_m = \frac{a^2}{2v_m}(t_a - t)^2 + v(t)$$

Then, the equation for curve B is

$$v(t) = v_m - \frac{a^2}{2v_m}(t_a - t)^2 \tag{2.20}$$

Distance traveled during the velocity curve B can be found by integrating Equation (2.20):

$$s_B(t) = s_A(t)|_0^{t_a/2} + \int_{t_a/2}^{t_a} (v_m - C_1(t^2 - 2t_a t + t_a^2))dt \tag{2.21}$$

where C_1 is the coefficient calculated in Equation (2.17). Also, the first term is the initial position for curve B which is found by evaluating Equation (2.19) at $t = t_a/2$.

S-CURVE VELOCITY PROFILE (FIGURE 2.9)

$$t_a = \frac{2v_m}{a}$$

$$C_1 = \frac{a^2}{2v_m}$$

Curve A

$0 \le t \le \frac{t_a}{2}$

$$s_A(t) = C_1 \frac{t^3}{3}$$

$$v_A(t) = C_1 t^2$$

$$a_A(t) = 2C_1 t$$

Curve B

$\frac{t_a}{2} < t \le t_a$

$$s_B(t) = C_1 \frac{t_a^3}{24} + v_m\left(t - \frac{t_a}{2}\right)$$

$$-C_1 \cdot \left\{ t_a^2\left(t - \frac{t_a}{2}\right) - t_a\left(t^2 - \left(\frac{t_a}{2}\right)^2\right) + \frac{1}{3}\left(t^3 - \left(\frac{t_a}{2}\right)^3\right) \right\}$$

$$v_B(t) = v_m - C_1(t_a - t)^2$$

$$a_B(t) = 2C_1(t_a - t)$$

■ EXAMPLE 2.2.5

An axis of a machine is to be moved with a pure S-curve velocity profile. Given the desired move velocity, $v_m = 10$ in/s and acceleration $a = 5$ in/s^2, what are the equations for velocity and acceleration during the curve A and curve B of the S-curve velocity profile?

Solution

Let us start with calculating the acceleration time:

$$t_a = 2\frac{v_m}{a} = \frac{2 \cdot 10}{5} = 4 \text{ s}$$

Calculating curve A of the S-curve profile from Equation (2.18) for $0 \le t \le \frac{t_a}{2}$ gives

$$v(t) = \frac{a^2}{2v_m}t^2 = \frac{5^2}{2 \cdot 10}t^2 = 1.25t^2$$

Differentiating this gives

$$a(t) = 2 \cdot (1.25t) = 2.5t$$

Calculating curve B of the S-curve profile from Equation (2.20) for $\frac{t_a}{2} < t \le t_a$ gives

$$v(t) = v_m - \frac{a^2}{2v_m}(t_a - t)^2$$

$$= 10 - \frac{5^2}{2 \cdot 10}(t^2 - 8t + 16)$$

$$= 10 - 1.25(t^2 - 8t + 16)$$

Differentiating this leads to

$$a(t) = 1.25(8 - 2t)$$

The velocity and acceleration profiles shown in Figure 2.10 were computed using these equations in MATLAB®.

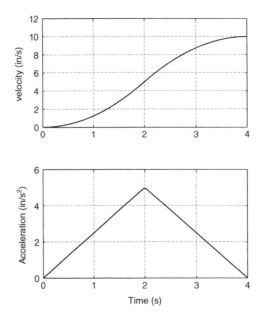

Figure 2.10 Pure S-curve velocity and acceleration profiles in Example 2.2.5

■ **EXAMPLE 2.2.6**

Usually axis position is measured using an encoder, which generates pulses as the axis moves. The number of pulses is proportional to the displacement of the axis. These pulses are counted by the motion controller to keep track of the axis position. Therefore, often motion profiles are programmed in counts (*cts*). Given the S-curve velocity profile in Figure 2.11, what is the axis position at $t = 100$ ms?

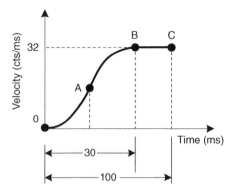

Figure 2.11   S-curve velocity profile for Example 2.2.6

Solution
We know that

$$t_a = \frac{2v_m}{a}$$

Solving for a:

$$a = \frac{2 \cdot 32}{30}$$
$$= 2.133 \, \text{cts/ms}^2$$

Distance traveled during $0 \leq t \leq \frac{t_a}{2}$ can be found from

$$s_{0A} = C_1 \frac{t^3}{3}$$

where $C_1 = a^2/(2v_m)$. Substituting the values gives $C_1 = 0.071$, and $s_{0A} = 80$ cts at $t = 15$.

Distance traveled during $\frac{t_a}{2} < t \leq t_a$ can be found from

$$s_{0B} = C_1 \frac{t_a^3}{24} + V_m \left(t - \frac{t_a}{2} \right)$$
$$- C_1 \cdot \left\{ t_a^2 \left(t - \frac{t_a}{2} \right) - t_a \left(t^2 - \left(\frac{t_a}{2} \right)^2 \right) + \frac{1}{3} \left(t^3 - \left(\frac{t_a}{2} \right)^3 \right) \right\}$$

Substituting $C_1 = 2.133, t_a = 15, t = 30, V_m = 32$ gives $s_{0B} = 480$ cts.

The distance traveled between points B and C can be found from the area of the rectangle under the velocity profile as

$$s_{BC} = 32 \cdot (100 - 30) = 2240 \, \text{cts}$$

Finally, the total distance traveled from points 0 to C is

$$s_{0C} = 480 + 2240 = 2720 \, \text{cts}$$

2.3   Multiaxis Motion

Multiaxes machines require coordination of the motion of individual axes to complete a task. Consider a CNC milling machine with a two-axis table and a vertical Z-axis with the spinning cutter. By coordinating the two axes of the table, we can create circular cuts in a part. Coordinating all three axes enables complex 3D cuts.

There are three basic approaches we can take in moving the axes of the machine [3]:

1. Move one axis at a time,
2. Start moving all axes at the same time (slew motion), and
3. Adjust the motion of the axes so that they all start and finish at the same time (interpolated motion).

2.3.1   Slew Motion

In slew motion, all axes start moving with the same speed and at the same time but each axis finishes its motion at a different time.

■ **EXAMPLE 2.3.1**

Consider the machine shown in Figure 2.12. If both axes are moving at the speed of 4 in/s using trapezoidal velocity profile with $t_a = 0.2$ s, how long will it take each axis to complete its move?

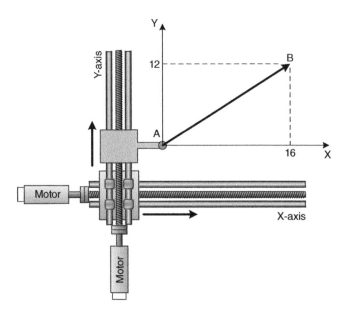

Figure 2.12   Multiaxis machine for Example 2.3.1 (top view)

Solution

The X-axis motion parameters are $t_a = 0.2$ s, $L = 16$ in, and $v_x = 4$ in/s. Using Equation (2.7), we can calculate the move time for this axis as

$$t_m^x = \frac{L}{v_m} - t_a$$

$$= \frac{16}{4} - 0.2$$

$$= 3.8 \text{ s}$$

The total time for the X-axis to complete its motion is $t_{\text{total}}^x = t_m^x + 2t_a$, which is 4.2 s. Similarly, the Y-axis move time is

$$t_m^y = \frac{L}{v_m} - t_a$$

$$= \frac{12}{4} - 0.2$$

$$= 2.8 \text{ s}$$

The total time for the Y-axis to complete its motion is $t_{\text{total}}^y = t_m^y + 2t_a$, which is 3.2 s. Hence, even though both axes start their motion at the same time, the Y-axis finishes its motion 1 s before the X-axis. As a result, the tool tip will not follow the straight line shown in Figure 2.12.

2.3.2   Interpolated Motion

In this mode, the motions of the axes are coordinated by the controller. For example, *linear* and *circular* interpolation can generate lines and circular segments, respectively, by coordinating the motion of two motors in a 2D plane. A third type of interpolated motion, called *contouring*, does not limit the number of motors or their motion profiles. It allows generation of any arbitrary trajectory in 3D space.

In the interpolated motion, faster axes are slowed down by the controller so that they finish at the same time as the axis that takes the longest time to complete its motion. There are two approaches to achieve this:

1. Slow down the faster axes while keeping the acceleration time, t_a, the same as the axis that takes the longest time to complete its motion
2. Slow down the faster axes while keeping the acceleration, a, the same as the axis that takes the longest time to complete its motion.

■ **EXAMPLE 2.3.2**

To make the tool tip in Example 2.3.1 follow the straight line between points "A" and "B", we can tell the controller to interpolate the motion. In this case, it will execute the motion of the longer move as programmed (X-axis) and slow down the shorter move (Y-axis) so that they both finish their moves at the same time.

Given $v_x = 4$ in/s and $t_a = 0.2$ s, what should be the new speed of the Y-axis, v_y, so that both axes finish their moves at the same time? Keep t_a the same for both axes.

Solution

From Example 2.3.1, we found that the X-axis will take 4.2 s to finish its move. Therefore, the total motion time for the Y-axis will also be $t^y_{total} = 4.2$ s. Recall that $t^y_{total} = t^y_m + 2t_a$. Then, $t^y_m = 3.8$ s. Rearranging Equation (2.7) gives

$$v_y = \frac{L}{(t_m + t_a)}$$

$$= \frac{12}{3.8 + 0.2}$$

$$= 3 \text{ in/s}$$

Since t_a is kept small, deviation from the straight line trajectory between points "A" and "B" will be very small.

Problems

1. A linear axis starts its motion from rest at position 0 with an acceleration of 2 in/s. After moving for 5 s, what is the position of the axis?
2. The shaft of a motor is rotating at 120 rpm. If it decelerates to stop in 2 s, how many revolutions will the shaft make during its deceleration?
3. Given the velocity profile in Figure 2.13, calculate s_A, s_B, s_C *using the geometry.*

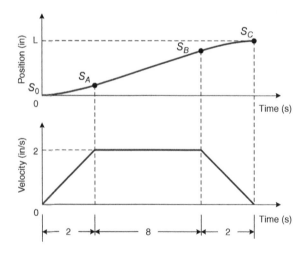

Figure 2.13 Problem 3

4. Given the same velocity profile in Problem 3, calculate s_A, s_B, s_C using Equation (2.4) in each segment of the profile.
5. In a motion control application, axis #1 needs to move 10 in with a speed of 2 in/s and acceleration of 4 in/s^2. What should be programmed into the motion controller for t_a and t_m in milliseconds (ms)?

6. Given the velocity profile in Figure 2.14 where all segments have equal time, derive the formula for v_m as a function of t and s_C.

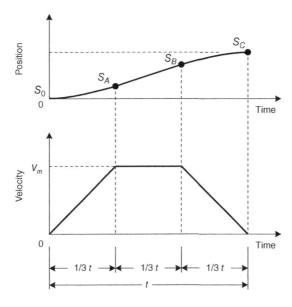

Figure 2.14   Problem 6

7. In the motion control application shown in Figure 2.15a, the nip rolls feed long wood boards which are cut to equal length pieces. Figure 2.15b shows the velocity profile for one cycle of the linear axis that carries the shear. The shear waits at a hover position over the conveyor. When it gets a signal from the controller, it accelerates and matches the speed of the conveyor (constant velocity segment). During this time, the shear is lowered and retracted to make the cut. After that, the axis slows down to zero speed and returns to the hover position. Find the return velocity V_{ret} to be programmed into the controller.

8. Motion controllers use *counts* (cts) to track position. The motor in Figure 2.16 has an encoder that produces 8000 cts/rev. The ball-screw advances 10 mm/rev. A motion application requires the carriage of the linear axis to travel 16 in at 4 in/s with an acceleration of 10 in/s^2. Calculate the following parameters for the carriage in the units shown: speed (cts/ms), distance (cts), acceleration (cts/ms^2), and move time (constant velocity time) in ms.

9. Given the S-curve velocity profile in Figure 2.17, plot the axis velocity and acceleration profiles as two separate plots using *Excel*®.

10. The X-axis of a XYZ gantry robot is going to operate using S-curve velocity profile with $V_m = 10$ in/s, $a = 5 \text{ in/s}^2$ for a total of 10 s. Compute and plot the resulting motion profile (trajectory, velocity, and acceleration) using MATLAB®. Use 100 time steps in each of the acceleration, maximum velocity, and deceleration regions.

11. When two consecutive moves with different speeds (V_1, V_2) are commanded as shown in Figure 2.18, the motion controller can blend the moves to create a smooth transition from

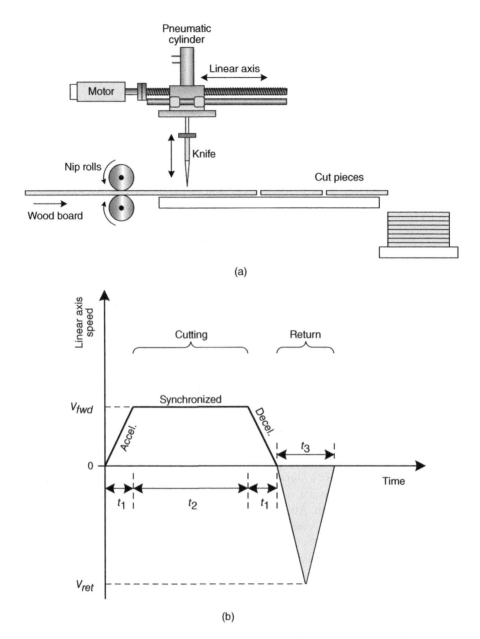

(a)

(b)

Figure 2.15 Problem 7. (a) Flying shear machine to cut continuous material into fixed lengths. (b) Velocity profile for the shear (Adapted by permission of ABB Corp.) [1]

one speed to the next. What is the equation of the blending velocity segment between points "A" and "B" in Figure 2.18? What are the accelerations a_1, a_2, and a_3?

12. Write a MATLAB® function to compute and plot a trapezoidal velocity profile and the corresponding position and acceleration profiles. Your function should take V_m, t_a, and t_m

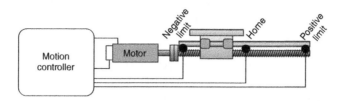

Figure 2.16 Problem 8

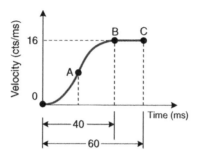

Figure 2.17 Problem 9

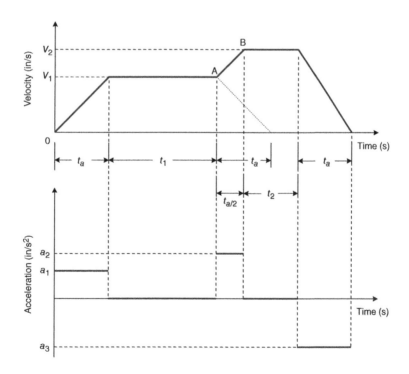

Figure 2.18 Problem 11

as inputs for the desired move. It should compute and return time, position, velocity, and acceleration in [t s v a] format. The MATLAB® function should have the following format:

```
function [t s v a] = TrapVelwithTM(Vm, ta, tm);
```

Try your function by calling it from the MATLAB® command prompt with $V_m = 2\,\text{in/s}, t_a = 2\,\text{s}$, and $t_m = 8\,\text{s}$ and verify the plot it generates.

13. In Example 2.3.2, the interpolated motion capability of a controller was shown. The speed of the Y-axis was slowed down by the controller so that both the X-axis and the Y-axis would finish their moves at the same time. The new speed for the Y-axis was calculated by keeping the acceleration time t_a the same for both axes.

Another way to slow down the faster axis is by *keeping its acceleration the same* as that of the slower axis. Given $V_x = 4$ in/s and $t_a = 0.2$ s, what should be the new speed V_y of the Y-axis so that both axes have the same acceleration and finish their moves at the same time?

References

[1] Application Note, Flying Shear (AN00116-003). Technical Report, ABB Corp., 2012.
[2] Delta Tau Data Systems Inc. *Application Note: Benefits of Using the S-curve Acceleration Component*, 2006.
[3] Hugh Jack (2013). Engineer On A Disk. http://claymore.engineer.gvsu.edu/~jackh/books/model/chapters /motion.pdf (accessed 5 August 2013).
[4] Chuck Lewin. *Mathematics of Motion Control Profiles*. PMD, 2007.

3

Drive-Train Design

Motion control systems are built using mechanical components that transmit motion from the actuator to the load (or the tool). The design problem typically requires selecting an appropriate motor and a transmission (such as belt drive or gearbox) so that the desired motion profile for the load can be achieved. The solution is found at the end of an iterative process. The motor and transmission combination is often called *drive-train* (Figure 3.1).

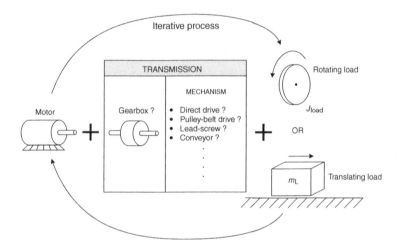

Figure 3.1 Iterative drive-train design process

There are four types of design problems we may encounter as listed in Table 3.1. However, the most common one is the first type where the desired load motion is specified and the motor and transmission need to be selected. The main goals of the design process are to

1. Ensure that torque available from motor (at maximum load speed) is greater than torque required by the application by a safety margin,
2. Ensure that proper inertia relationship between the motor and the load is met, and
3. Meet any additional criteria (cost, precision, stiffness, cycle time, etc.).

Industrial Motion Control: Motor Selection, Drives, Controller Tuning, Applications, First Edition. Hakan Gürocak.
© 2016 John Wiley & Sons, Ltd. Published 2016 by John Wiley & Sons, Ltd.

Table 3.1 Types of drive-train design problems

Type	Given	Find/size
1	Desired load motion	Transmission and motor
2	Existing motor and transmission	Resulting load motion
3	Existing motor, desired load motion	Transmission
4	Desired load motion, transmission	Motor

Motor selection is often called *motor sizing* in motion control industry. The motor size refers to the torque and power of a motor. Oversizing a motor by a large margin increases the cost of the system and will lead to slower system response as most of the energy will be spent in accelerating the motor inertia. An undersized motor will not be able to provide the necessary load motion. In certain situations, even if it barely meets the load motion requirements, it will most likely have a short life due to overheating.

This chapter starts by introducing the concept of inertia and torque reflection. Then, the concept of inertia ratio is presented. Next, five types of transmission mechanisms are explored in depth in terms of the transmission ratio, inertia, and torque reflection in each case. Motor sizing requires analysis of the torque requirements for the desired motion profile. After introducing the torque–speed curves for three-phase AC servo motors and induction motors, motor selection procedures are presented for direct-drive and for axes with transmission mechanisms. Planetary servo gearhead for servo motors and worm gear speed reducers for vector-duty induction motors are also explained. chapter concludes by providing selection procedures for axes with motor, gearbox, and transmission mechanisms.

3.1   Inertia and Torque Reflection

The mass moment of inertia, J, is a property of an object. It combines the mass and the shape of the object into a single quantity. In this book, the term inertia is used in short to refer to the mass moment of inertia.

Inertia defines the resistance of an object to change in its angular velocity about an axis of rotation. In other words, inertia opposes change in motion. In rotational dynamics, Newton's second law is

$$\sum T = J\alpha \tag{3.1}$$

where T is torque and α is angular acceleration. Comparing this to the traditional form of Newton's second law $\left(\sum F = ma \right)$, we can see that inertia is the rotational equivalent of mass in linear motion. Due to this analogy, both rotating and translating masses are simply referred to as inertia in drive-train design.

3.1.1   Gearbox Ratio

Gearbox ratio is defined as

$$N_{\text{GB}} = \frac{\text{motor speed}}{\text{load speed}} \tag{3.2}$$

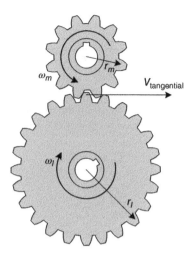

Figure 3.2 Gear mesh

Often a notation such as 5:1 is used to designate the gearbox ratio. This means $N_{GB} = 5$, where the motor speed is 5 times faster than the load speed. Parameters other than the shaft speeds can also be used to define the gearbox ratio.

Tangential Speed

At the mesh point between the gears in Figure 3.2, we can write the tangential speed as

$$V_{\text{tangential}} = \omega_m r_m = \omega_l r_l$$

where ω_m is the speed of the motor gear (or shaft) in rad/s, ω_l is the speed of the load gear (or shaft) in rad/s, r_m and r_l are the pitch circle radii of the motor and load gears, respectively, in inches or mm. We can rewrite this equation as

$$\frac{\omega_m}{\omega_l} = \frac{r_l}{r_m} \tag{3.3}$$

or, using Equations (3.2) and (3.3),

$$N_{GB} = \frac{\omega_m}{\omega_l} = \frac{r_l}{r_m} \tag{3.4}$$

Number of Gear Teeth

Another way of deriving the gear ratio is by using the number of teeth on each gear. The number of teeth on a gear is directly proportional to its size (diameter or radius). For example, if one of the two meshing gears is larger than the other one, the larger gear will have more teeth on it. Therefore,

$$\frac{n_l}{n_m} = \frac{r_l}{r_m}$$

where n_l and n_m are the number of teeth on the load and motor gears, respectively.

Equation (3.4) can now be rewritten as

$$N_{GB} = \frac{\omega_m}{\omega_l} = \frac{r_l}{r_m} = \frac{n_l}{n_m} \tag{3.5}$$

Torque

Yet another way to define the gear ratio is by using the torques on the input and output shafts of the gear drive. Assuming 100% efficiency, power ($\mathbb{P}$) transmitted through the gear drive remains constant. Therefore,

$$\mathbb{P} = T_m \omega_m = T_l \omega_l \tag{3.6}$$

or,

$$\frac{\omega_m}{\omega_l} = \frac{T_l}{T_m}$$

where T_l and T_m are the torque on the load gear (or shaft) and the motor gear (or shaft), respectively. Then, Equation (3.5) becomes

$$N_{GB} = \frac{\omega_m}{\omega_l} = \frac{r_l}{r_m} = \frac{n_l}{n_m} = \frac{T_l}{T_m} \tag{3.7}$$

3.1.2   Reflected Inertia

The inertia seen (or felt) by the motor changes when the load is coupled to the motor through a gearbox as opposed to direct coupling. For the directly coupled load in Figure 3.3a, the equation of motion for the load can simply be written as

$$T_m = J_{load} \ddot{\theta}_m$$

When the same load is coupled to the motor through a gearbox as shown in Figure 3.3b, neglecting the gear inertia we can write the equation of motion for the load as

$$T_l = J_{load} \ddot{\theta}_l$$

where T_l is the motor torque transferred to the load through the gears and $\ddot{\theta}_l$ is the angular acceleration of the load shaft. Using Equation (3.7),

$$\frac{r_l}{r_m} T_m = J_{load} \ddot{\theta}_l \tag{3.8}$$

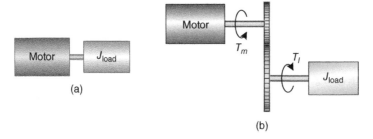

(a)

(b)

Figure 3.3   Inertia reflection. (a) Load directly coupled to motor. (b) Load coupled to motor through gears

When the gears rotate, the distance traveled along the circumference of each gear is the same and given by

$$r_l \theta_l = r_m \theta_m$$

If we differentiate both sides twice, we can also write

$$r_l \ddot{\theta}_l = r_m \ddot{\theta}_m \tag{3.9}$$

Solving for $\ddot{\theta}_l$ from Equation (3.9) and substituting into (3.8) gives

$$\frac{r_l}{r_m} T_m = J_{\text{load}} \frac{r_m}{r_l} \ddot{\theta}_m$$

or,

$$T_m = J_{\text{load}} \left(\frac{r_m}{r_l} \right)^2 \ddot{\theta}_m$$

$$= J_{\text{load}} \frac{1}{N_{\text{GB}}^2} \ddot{\theta}_m \tag{3.10}$$

Comparing this result to Equation (3.1), we can see that Equation (3.10) is Newton's second law applied to the motor shaft. Hence, the term in front of $\ddot{\theta}_m$ must be the load inertia reflected to the motor through the gears

$$J_{\text{ref}} = \frac{J_{\text{load}}}{N_{\text{GB}}^2} \tag{3.11}$$

Speed-changing elements, such as gears or pulleys, follow this relationship to modify the load inertia seen by the motor.

3.1.3 Reflected Torque

Going back to Equation (3.6), we can write

$$T_m = \frac{\omega_l}{\omega_m} T_l$$

$$= \frac{T_l}{N_{\text{GB}}} \tag{3.12}$$

Note that, unlike in the inertia reflection, the gear ratio N_{GB} is *not* squared. Speed-changing elements also modify the torque reflected to the motor.

3.1.4 Efficiency

Equations (3.11) and (3.12) are for *ideal* motion transmission where no energy is lost. In other words, the mechanism is 100% efficient. However, in a real gear drive, *efficiency is always less than 100%* since some of the input power is lost through the transmission due to friction and heating. The efficiency, η, of a transmission is defined as the ratio of the output power to the input power as

$$\eta = \frac{\mathbb{P}_{\text{output}}}{\mathbb{P}_{\text{input}}} \tag{3.13}$$

From Equations (3.6) and (3.13),

$$T_l \omega_l = \eta \, T_m \omega_m$$

Then, we can finalize Equation (3.12) to incorporate the efficiency

$$T_m = \frac{T_l}{\eta N_{\text{GB}}} \tag{3.14}$$

Similarly, Equation (3.11) can be finalized to include efficiency

$$J_{\text{ref}} = \frac{J_{\text{load}}}{\eta N_{\text{GB}}^2} \tag{3.15}$$

In designing motion control systems, one approach is to use Equations (3.11) and (3.12) throughout the calculations assuming 100% efficiency and then account for the losses by using a safety factor in determining the size of actuators at the end. Another approach is to explicitly account for the efficiency in the calculations by using Equations (3.14) and (3.15). In some of the manufacturer catalogs, Equations (3.11) and (3.14) are used instead.

3.1.5   Total Inertia

If a gearbox or transmission mechanism (such as pulley-and-belt) is used in the design of an axis, some inertia will be on the motor shaft while others will be on the load shaft. There is also the inertia of the motor J_m (rotor inertia).

A convenient way to properly account for all inertia in the system is to *reflect* all inertia to the motor shaft. Then, the total inertia on the motor shaft consists of the following

---

TOTAL INERTIA ON MOTOR SHAFT

$$J_{\text{total}} = J_m + J_{\text{on motor shaft}} + J_{\text{ref}} \tag{3.16}$$

---

where $J_{\text{on motor shaft}}$ on motor shaft is the total *external* inertia on the motor shaft. J_{ref} is the total inertia reflected to the motor shaft.

■ EXAMPLE 3.1.1

Given the system below, find the equivalent total inertia J_{total} on the motor shaft.

Solution
The total inertia is given by Equation (3.16). In this system, the gear inertia are taken into account. The coupling inertia and motor gear inertia are on the motor shaft. Therefore,

$$J_{\text{on motor shaft}} = J_{\text{coupling}} + J_{mg}$$

The load inertia and the load gear inertia are on the load shaft. Therefore, they need to be first reflected to the motor shaft using Equation (3.15) before they can be added to the rest of the inertia

$$J_{\text{ref}} = \frac{1}{\eta N_{\text{GB}}^2} \left[J_{lg} + J_{\text{load}} \right] \tag{3.17}$$

Then, the total inertia on the motor shaft becomes

$$J_{\text{total}} = J_m + J_{\text{coupling}} + J_{mg} + \frac{1}{\eta N_{\text{GB}}^2} \left[J_{lg} + J_{\text{load}} \right] \tag{3.18}$$

In this example, we ignored the shaft inertia. If necessary, they can also be included in these calculations.

We can draw the same system in Figure 3.4 as a schematic shown in Figure 3.5. Comparing this figure to Equation (3.18) shows the one-to-one match between the elements in the figure and the individual terms in the equation. Note that the total inertia on the output side (shaft #2) of the gearbox is reflected to the input side (shaft #1) of the gearbox by the $\frac{1}{\eta N_{\text{GB}}^2}$ multiplier, which is shown by the gearbox in the schematic.

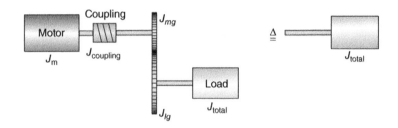

Figure 3.4   Inertia reflection through gears and the equivalent system

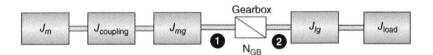

Figure 3.5   Another schematic for the system in Figure 3.4

3.2   Inertia Ratio

Inertia ratio, J_R, is defined as

$$J_R = \frac{J_{\text{on motor shaft}} + J_{\text{ref}}}{J_m}$$

where the numerator consists of the sum of all inertia *external* to the motor. Then, the inertia ratio simply compares the total load inertia the motor has to work against to the inertia of the motor itself.

Consider the system in Example 3.1.1 where a simple gearbox connects the load to the motor. As it can be seen from Equations (3.17) and (3.18), the inertia of the motor and load gears (J_{mg}, J_{lg}) appear in both $J_{\text{on motor shaft}}$ and J_{ref}. In a simple gearbox like this, it is easy to calculate the individual gear inertia and properly include them in finding the inertia ratio. However, often a commercially available gearbox is selected from a catalog to be used in the design of the system. In this situation, the internal design details of the gearbox will be unknown making it difficult to properly calculate the inertia ratio. Fortunately, gearbox manufacturers provide the gearbox inertia *as reflected to the input shaft* (motor shaft) of the gearbox. Then, the inertia ratio can be found from

$$J_R = \frac{J_{\text{on motor shaft}} + J_{\text{load}\rightarrow M} + J_{\text{GB}\rightarrow M}}{J_m} \tag{3.19}$$

where $J_{\text{GB}\rightarrow M}$ is the inertia of the gearbox *reflected to its input (motor shaft)*, J_m is the motor inertia, and $J_{\text{load}\rightarrow M}$ is the load inertia reflected to the motor shaft.

■ **EXAMPLE 3.2.1**

The system in Figure 3.6 uses a PN023 gearbox [5] by Apex Dynamics, Inc. It has 5:1 gear ratio, $0.15 \, \text{kg-cm}^2$ inertia reflected to the input side and 97% efficiency. The motor is a Quantum QB02301 NEMA size 23 servomotor [2] by Allied Motion Technologies, Inc. It has $1.5 \times 10^{-5} \, \text{kg-m}^2$ rotor inertia. If the load inertia is $10 \times 10^{-4} \, \text{kg-m}^2$, find the inertia ratio.

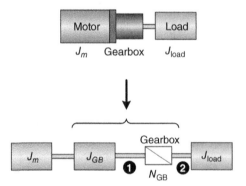

Figure 3.6   Inertia reflection through a gearbox

Solution
Figure 3.6 shows a schematic for this system. Since the gearbox inertia given by the manufacturer is already reflected to the motor side, the schematic has it on the motor shaft. The load inertia reflected to the motor shaft is

$$J_{\text{load}\rightarrow M} = \frac{J_{\text{load}}}{\eta N_{\text{GB}}^2}$$

$$= \frac{10 \times 10^{-4}}{0.97 \cdot 5^2}$$

$$= 4.124 \times 10^{-5} \, \text{kg-m}^2$$

From Equation (3.19), the inertia ratio is found as

$$J_R = \frac{J_{\text{on motor shaft}} + J_{\text{load}\to M} + J_{GB\to M}}{J_m}$$

$$= \frac{4.124 \times 10^{-5} + 0.15 \times 10^{-4}}{1.5 \times 10^{-5}}$$

$$= 3.75$$

where $J_{\text{on motor shaft}} = 0$ since there is no other inertia on the motor shaft (ignoring the shafts).

3.2.1   Targeted Practical Inertia Ratio

Performance requirements of an application must be clearly understood and the inertia ratio needs to be selected accordingly. Several motors may be available that would provide the necessary inertia ratio. The design task is to find the smallest motor that can produce the speed and torque required for the application.

For practical purposes, the inertia ratio should be [11, 25]

$$J_R \leq 5 \tag{3.20}$$

Performance tends to go up as the inertia ratio is lowered. If the machine needs to be nimble with fast moves and frequent starts and stops, inertia ratios down to 2 or 1 can be used [11]. But if high performance and fast dynamics are not critical, inertia ratio of 10 is common and, 100, or even higher are possible.

In general, ease of controller tuning and machine performance go up as the inertia ratio goes down. If all other factors are equal, lower inertia ratio is better. However, if the inertia ratio is very low, the motor may be oversized, and therefore expensive and bulky, with not much improvement in overall machine performance.

The choice of the inertia ratio also depends on the stiffness of the system. A system is considered stiff if it does not deflect, stretch, or bend under the load. A system with a motor and gearbox that are properly connected to the load is considered stiff. For a stiff system, the inertia ratio can be 5–10. On the other hand, systems using belt and pulleys may flex under load due to the stretching of the belt. Hence, the inertia ratio should be kept small.

3.3   Transmission Mechanisms

In Section 3.1.2, we looked at the case where the load was coupled to the motor through a gearbox. Most mechanical systems involve transmission mechanisms between the load and the motor. A transmission mechanism connects the load to the motor and helps meet the motion profile requirements. In this section, we will look at the details of five types of transmission mechanisms:

- Pulley-and-belt
- Lead screw and ball screw

- Rack-and-pinion
- Belt drive for linear motion
- Conveyor.

Except for the pulley-and-belt, all the other mechanisms convert rotational motion into linear motion.

3.3.1 Load and Inertia Reflection through Transmission Mechanisms

In Section 3.1.2, we saw that speed-changing elements, such as gearboxes, modify the inertia and torque seen by the motor. Just like the gearboxes, transmission mechanisms also modify the inertia and torque seen by the motor. Each transmission mechanism has a transmission ratio "N", similar to the gearbox ratio N_{GB}.

Figure 3.7 is the schematic of a typical drive-train with a transmission mechanism T where the load is either rotating or translating. In this configuration, the total inertia seen by the motor is given by

$$J_{total} = J_m + J_{C1} + J_{ref}^{trans} \tag{3.21}$$

where J_m is the motor inertia, J_{C1} is the inertia of the motor coupler, and J_{ref}^{trans} is the inertia reflected by the transmission mechanism to its input shaft. Each transmission mechanism T has a specific formula for the J_{ref}^{trans} term in this equation.

External torques or forces acting on the load are reflected to the input shaft of the transmission mechanism as torque demand, $T_{load \rightarrow in}$, from the motor (Figure 3.8). Each transmission mechanism reflects the load torque demand to its input shaft in a specific way based on its kinematics.

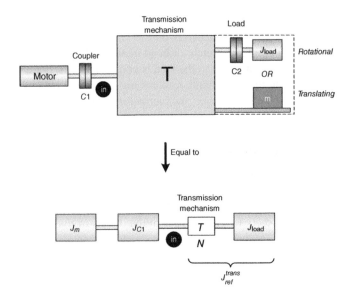

Figure 3.7   Schematic of a typical drive with a transmission mechanism

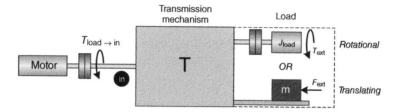

Figure 3.8 External forces or torques on the load are reflected by the transmission as a torque demand $T_{\text{load}\to\text{in}}$ on the motor

3.3.2 Pulley-and-Belt

A pulley-and-belt transmission consists of two pulleys and a belt connecting them. In motion control systems, belts with teeth (timing belts) are used as shown in Figure 3.9. This enables more accurate positioning of the load with no slippage of the belt. The pulleys for timing belts are called sprockets.

Transmission Ratio

In Figure 3.10, the linear speed of a point on the belt can be written in terms of the angular speeds of each pulley as

$$V_{\text{tangential}} = \omega_{ip} \cdot r_{ip} = \omega_{lp} \cdot r_{lp}$$

where ω_{ip} is the speed of the input pulley in rad/s, ω_{lp} is the speed of the load pulley in rad/s, r_{ip} and r_{lp} are the radii of the input and load pulleys, respectively.

Figure 3.9 Pulley-and-belt transmission using timing belt and sprockets (Reproduced by permission of The Gates Corp. [35])

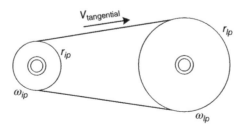

Figure 3.10 Pulley-and-belt drive

Then, we can rewrite this equation to define the transmission ratio as

$$N_{BP} = \frac{\omega_{ip}}{\omega_{lp}} = \frac{r_{lp}}{r_{ip}} \tag{3.22}$$

Reflected Inertia

Figure 3.11a shows the top view of belt-and-pulley transmission. The same drive-train is shown as a schematic in Figure 3.11b, which corresponds to the schematic in Figure 3.8.

The inertia reflected by the mechanism to its input shaft is given by

$$
\begin{aligned}
J_{ref}^{trans} &= J_{IP} + J_{belt \to in} + J_{LP \to in} + J_{load \to in} + J_{C2 \to in} \\
&= J_{IP} + \left(\frac{W_{belt}}{g \cdot \eta} \right) \cdot r_{ip}^2 + \frac{1}{\eta N_{BP}^2} (J_{LP} + J_{load} + J_{C2})
\end{aligned} \tag{3.23}
$$

where J_{IP} is the inertia of the input pulley, J_{load} is the inertia of the load, J_{LP} is the inertia of load pulley, and J_{C2} is the inertia of the load coupler.

The $J_{belt \to in}$ term models the belt as a spinning mass m whose inertia is given by

$$J = mr^2$$

Substituting $m = W_{belt}/g$ and $r = r_{ip}$ gives the $J_{belt \to in}$ term in Equation (3.23) where W_{belt} is the weight of the belt, g is gravitational acceleration, and r_{ip} is the radius of the input pulley.

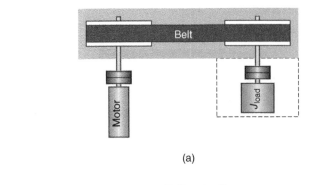

(a)

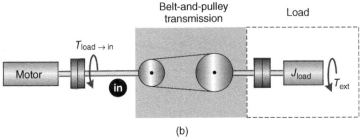

(b)

Figure 3.11   Belt-and-pulley transmission. (a) Top view. (b) Schematic

Load Torque

Figure 3.11b shows the total external torque T_{ext} acting on the load and the resulting torque demand $T_{load \rightarrow in}$ from the motor reflected through the belt-and-pulley transmission. As in the case of the gearbox, the load torque seen by the motor is given by Equation (3.14). To apply this equation to the belt-and-pulley transmission, we need to calculate the transmission ratio from Equation (3.22) and account for the mechanism efficiency η. Therefore, Equation (3.14) is rewritten as

$$T_{load \rightarrow in} = \frac{T_{ext}}{\eta N_{BP}}$$

where T_{ext} is the summation of all external torques acting on the load given by Equation (3.36) explained later in Section 3.4.

3.3.3   Lead Screw

Lead screws are widely used in converting rotary motion into linear motion. Two most common types of lead screws use *ACME screw* and *ball screw* as shown in Figures 3.12a and 3.12b, respectively. The ACME screw is difficult to back drive. In other words, motor can drive the load but the load cannot drive the motor (e.g., when the motor power is turned off in a vertical axis). The ACME screws can transmit large forces and therefore often are called power screws. Efficiency of an ACME screw can be in the range of 35–85%. The ball-screw uses precision ground ball bearings in a groove. The screw and the nut do not touch each other. Ball bearings in the grooves between the screw and the nut recirculate as the screw (or nut) rotates. When the balls reach trailing end of the nut, they are guided by a return tube back to the leading end of the nut and continue to recirculate [34]. Reduced backlash and friction makes ball-screws very popular in motion control applications. Efficiency of the ball-screw is in the range of 85–95%. Since the output of the mechanism is linear motion, torque input from the motor is converted into force at the output.

Transmission Ratio

The transmission ratio of the lead screw can be calculated from the definition of pitch [23].

Pitch (rev/in): Number of screw revolutions required for the nut to travel 1 in (or rev/m in SI units).

Lead (in/rev): Distance traveled by the nut in one revolution of the screw (m/rev in SI units).

(a)   (b)

Figure 3.12   ACME screw and ball-screw used in lead screw transmission (Reproduced by permission of Rockford Ball Screw Co.). (a) ACME screw [33]. (b) Ball screw [34]

Then, we can express the definition of pitch as an equation

$$\Delta\theta = 2\pi p \Delta x \qquad (3.24)$$

where $\Delta\theta$ is the rotation of input shaft (rad), p is pitch, and Δx is the linear displacement of the nut (m or in).

In Equation (3.2), the transmission ratio was defined as the motor speed divided by the load speed. Then, assuming we divide both sides of Equation (3.24) by time Δt, we can rearrange it as

$$\frac{\text{input speed}}{\text{load speed}} = \frac{\dot{\theta}}{\dot{x}} = 2\pi p \qquad (3.25)$$

Hence, the transmission ratio for the lead screw (or ball-screw) mechanism is

$$N_S = 2\pi p$$

Reflected Inertia

We need to first derive the relationship between the inertia of the translating mass and reflected inertia. The kinetic energy of the total mass m in linear motion is

$$KE = \frac{1}{2}m\dot{x}^2$$

Using Equation (3.25), we can rewrite the kinetic energy as

$$KE = \frac{1}{2}m\frac{1}{(2\pi p)^2}\dot{\theta}^2$$

Since the speed is now expressed as angular speed, the term in front of it must then be equal to the reflected inertia

$$J_{ref} = m\frac{1}{(2\pi p)^2}$$

or,

$$J_{ref} = \frac{m}{N_S^2}$$

This result is similar to what we had in Equation (3.15). The load inertia is now replaced by the total mass in linear motion and the gearbox ratio is replaced by the transmission ratio N_S for the screw. The total mass in linear motion can be found from the weights of the load (W_L) and the carriage (W_C) as

$$m = \frac{W_L + W_C}{g}$$

Figure 3.13a and b shows the schematic of a drive-train with screw and a commercial product, respectively.

The inertia reflected by the mechanism to its input shaft is given by

$$J_{ref}^{trans} = J_{screw} + J_{load\to in} + J_{carriage\to in}$$

$$= J_{screw} + \frac{1}{\eta N_S^2}\left(\frac{W_L + W_C}{g}\right) \qquad (3.26)$$

where J_{screw} is the inertia of the screw.

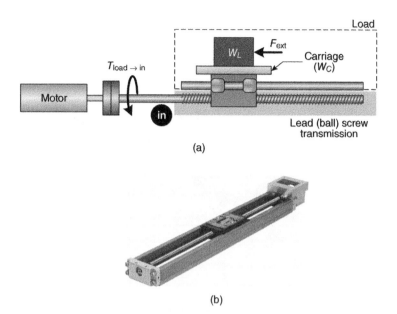

(a)

(b)

Figure 3.13   Lead (ball) screw transmission. (a) Schematic. (b) Electric slide with ball screw (Reproduced by permission of Festo Corp.) [14]

Load Torque

The nut on the screw works against all external forces, F_{ext} given by

$$F_{ext} = F_f + F_g + F_p \tag{3.27}$$

where F_f is the friction force, F_g is the gravitational force component *along the screw axis*, and F_p is an external force acting on the carriage due to the interaction of the mechanism with the environment (such as forces on the tool of a machine during assembly).

To study these forces, consider the screw transmission in Figure 3.14 where the mechanism is at an angle β with the horizontal. The total weight in motion is the summation of the load and carriage weights. We can express the friction force and the gravitational force component as

$$F_f = \mu (W_L + W_C) \cos \beta$$
$$F_g = (W_L + W_C) \sin \beta \tag{3.28}$$

where μ is the friction coefficient for the screw. Then, we can rewrite Equation (3.27) as

$$F_{ext} = F_p + (W_L + W_C)(\sin \beta + \mu \cos \beta) \tag{3.29}$$

Note that when the mechanism is horizontal ($\beta = 0$), the gravitational force component F_g that the mechanism must work against becomes zero.

The load torque as seen by the motor can be calculated by considering the work done [24]

$$\text{Work} = F_{ext} \Delta x \tag{3.30}$$

where Δx is the linear distance traveled by the load.

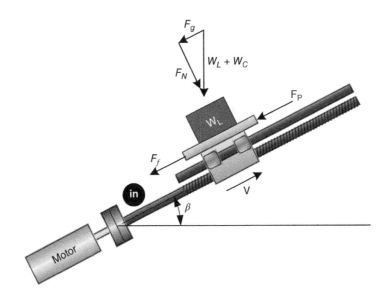

Figure 3.14   Forces on a lead screw transmission at an angle with the horizontal

Using Equation (3.24) in (3.30)

$$\text{Work} = F_{\text{ext}} \frac{1}{(2\pi p)} \Delta\theta$$

but also from the input side

$$\text{Work} = T_{\text{load}\rightarrow\text{in}} \Delta\theta$$

Hence,

$$T_{\text{load}\rightarrow\text{in}} = \frac{F_{\text{ext}}}{N_S}$$

accounting for the drive-train efficiency (η),

$$T_{\text{load}\rightarrow\text{in}} = \frac{F_{\text{ext}}}{\eta N_S}$$

This result is very similar to what we had in Equation (3.14) with the total external torque now replaced by the sum of all external forces and the gearbox ratio replaced by the transmission ratio for the screw.

■ **EXAMPLE 3.3.1**

A 100 lb load will be positioned using a steel ball-screw with 0.28 lb/in^3 density, 0.375 in diameter, 36 in length, 0.75 in/rev lead, and 90% efficiency. The carriage weighs 0.47 lb. Calculate the inertia reflected by the transmission to its input shaft.

Solution

From Equation (3.26), the reflected inertia is given by

$$J_{ref}^{trans} = J_{screw} + J_{load \to in} + J_{carriage \to in}$$

$$= J_{screw} + \frac{1}{\eta N_S^2} \left(\frac{W_L + W_C}{g} \right)$$

where $W_L = 100\,lb$, $W_C = 0.47\,lb$, and $g = 386\,in/s^2$. The transmission ratio of the ball-screw is given by $N_s = 2\pi p$ where p is the pitch of the screw and is equal to the inverse of its lead specified in the problem. Then,

$$N_s = 2\pi p$$

$$= 2\pi \frac{1}{0.75} = 8.38$$

The screw inertia J_{screw} can be calculated by approximating the screw as a long slender solid cylinder

$$J_{screw} = \frac{\pi L \rho D^4}{32g}$$

$$= \frac{\pi \cdot 36 \cdot 0.28 \cdot 0.375^4}{32 \cdot 386}$$

$$= 5.07 \times 10^{-5}\,lb\text{-}in\text{-}s^2$$

where D is the pitch diameter and L is the length of the screw in inches. A very important point to remember in using this formula is the unit of the material density ρ. In the U.S. customary units, the material density is specified as lb/in^3, which is called weight density. Therefore, the formula contains the gravitational acceleration g in the denominator to convert the weight density into mass density.

Often the same formula is specified as

$$J_{screw} = \frac{\pi L \rho D^4}{32}$$

where ρ must be mass density and the formula does not contain the gravity term g. In the International System of Units (SI units), the density is specified as kg/m^3, which is mass density. Therefore, the formula without the gravity term needs to be used when calculating the inertia in SI units.

The inertia reflected by the transmission to its input shaft is calculated as

$$J_{ref}^{trans} = J_{screw} + \frac{1}{\eta N_S^2} \left(\frac{W_L + W_C}{g} \right)$$

$$= 5.07 \times 10^{-5} + \frac{1}{0.9 \cdot 8.38^2} \left(\frac{100 + 0.47}{386} \right)$$

$$= 4.17 \times 10^{-3}\,lb\text{-}in\text{-}s^2$$

3.3.4   Rack-and-Pinion Drive

Rack-and-pinion in Figure 3.15 is another common type of transmission used to convert rotary motion into linear motion.

Transmission Ratio

The transmission ratio between the linear motion of the load and the rotational motion of the pinion can be calculated from the following relationship

$$V_{rack} = r_{pinion}\, \omega_{inion}$$

Using the definition of gear ratio from Equation (3.2), the transmission ratio for the rack-and-pinion can be written as

$$N_{RP} = \frac{1}{r_{pinion}}$$

NOTE: This equation is valid if the angular speed of the pinion is in rad/s.

Reflected Inertia

Figure 3.16 shows the schematic of a drive-train with rack-and-pinion. The inertia reflected by the mechanism to its input shaft is given by

$$J_{ref}^{trans} = J_{pinion} + J_{load \to in} + J_{carriage \to in}$$

$$= J_{pinion} + \frac{1}{\eta N_{RP}^2}\left(\frac{W_L + W_C}{g}\right)$$

where J_{pinion} is the inertia of the pinion gear. Both the carriage and load are translating masses.

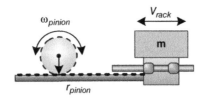

Figure 3.15   Rack-and-pinion drive converts rotational motion into linear motion

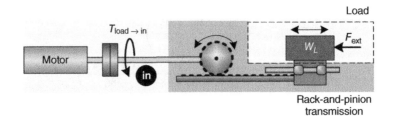

Figure 3.16   Schematic for rack-and-pinion transmission

Load Torque

Just like the lead screw transmission, the total external force acting on the mechanism along the motion direction of the load can be found from Equation (3.29). Then, the torque demand reflected to the input shaft of the pinion is

$$T_{\text{load}\to\text{in}} = \frac{F_{\text{ext}}}{\eta N_{\text{RP}}}$$

3.3.5   Belt-Drive for Linear Motion

If the load is connected to a belt and two identical pulleys as shown in Figure 3.17, then the rotary motion of the motor can be converted into linear motion of the load. This is a drive-train frequently used in low-inertia and low-load motion control applications.

Transmission Ratio

The transmission ratio between the linear motion of the load and the rotational motion of the input shaft of the belt-drive can be derived using the same approach as in the rack-and-pinion. The pinion radius is replaced by the input pulley radius, r_{ip}

$$N_{\text{BD}} = \frac{1}{r_{ip}} \tag{3.31}$$

Again, this equation is valid if the angular speed of the motor pulleys is in rad/s.

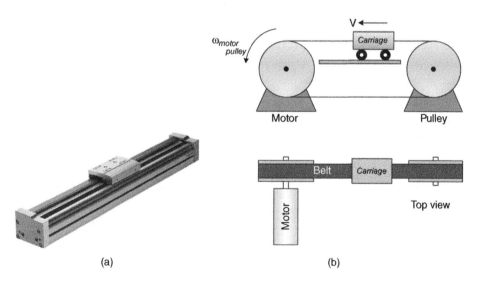

(a)   (b)

Figure 3.17   Belt-drive for linear motion converts rotational motion of the motor into linear motion of the load. (a) Electric linear axis (Reproduced by permission of Festo Corp.) [13]. (b) Kinematics of belt drive for linear motion

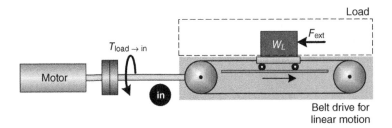

Figure 3.18　Schematic for belt-drive for linear motion

Reflected Inertia

Figure 3.18 shows the schematic of a belt-drive for linear motion. The inertia reflected by the mechanism to its input shaft is given by

$$J_{\text{ref}}^{\text{trans}} = J_{\text{IP}} + J_{\text{load}\to\text{in}} + J_{\text{carriage}\to\text{in}} + J_{\text{belt}\to\text{in}} + J_{\text{LP}}$$

$$= 2J_P + \frac{1}{\eta N_{\text{BD}}^2}\left(\frac{W_L + W_C + W_{\text{belt}}}{g}\right) \tag{3.32}$$

where $J_P = J_{\text{IP}} = J_{\text{LP}}$ are the inertia of the two pulleys. Since they are identical, the inertia of the load pulley appears directly as if it were on the input shaft. In other words, the transmission ratio from the load pulley to the input pulley through the belt is equal to one.

Load Torque

Just like the lead screw transmission, the friction force and gravitational force component along the motion direction of the load can be found from Equation (3.28). Then, the torque demand reflected to the input pulley is

$$T_{\text{load}\to\text{in}} = \frac{F_{\text{ext}}}{\eta N_{\text{BD}}} \tag{3.33}$$

3.3.6　Conveyor

A conveyor can have one or more idler rollers as shown in Figure 3.19. This enables the use of longer belts and handling heavier loads.

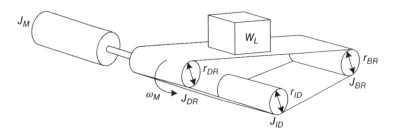

Figure 3.19　Conveyor

Transmission Ratio

As in the case of the rack-and-pinion, a transmission ratio between the linear motion of the load on the belt and the rotational motion of the drive roller (DR) of the conveyor can be calculated from the following relationship

$$N_{CV} = \frac{1}{r_{DR}}$$

Reflected Inertia

Figure 3.20 shows the schematic of a conveyor. The inertia reflected by the mechanism to its input shaft is given by

$$J_{ref}^{trans} = J_{DR} + J_{load \to in} + J_{belt \to in} + J_{ID \to in} + J_{BR \to in}$$

$$= J_{DR} + \frac{1}{\eta N_{CV}^2} \left(\frac{W_L + W_{belt}}{g} \right) + \frac{J_{ID}}{\eta \left(\frac{r_{ID}}{r_{DR}} \right)^2} + \frac{J_{BR}}{\eta \left(\frac{r_{BR}}{r_{DR}} \right)^2}$$

where J_{DR} is the inertia of the drive roller, J_{ID} and J_{BR} are the inertia of the idler roller and the back roller, respectively.

As shown in Figure 3.19, the belt connects the drive roller on the input shaft to the idler, ID as if they were two gears in mesh. Therefore, from Equation (3.15),

$$J_{ID \to in} = \frac{J_{ID}}{\eta \left(\frac{r_{ID}}{r_{DR}} \right)^2}$$

where the squared term in the denominator is the transmission ratio between the drive roller DR and the idler ID from Equation (3.22).

Similarly, the belt connects the drive roller *DR* on the input shaft to the back roller *BR* as if they were two gears in mesh. Therefore, again from Equation (3.15),

$$J_{BR \to in} = \frac{J_{BR}}{\eta \left(\frac{r_{BR}}{r_{DR}} \right)^2}$$

Load Torque

Same equation as in the lead screw case also applies here as the load translates while the belt pulleys spin. Considering a general case where the top of the conveyor belt has an angle β with

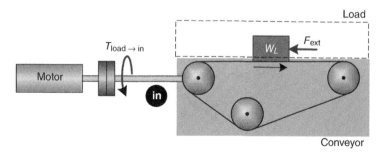

Figure 3.20   Schematic for conveyor

the horizontal, the torque demand reflected to the input pulley shaft is

$$T_{\text{load}\rightarrow\text{in}} = \frac{F_{\text{ext}}}{\eta N_{\text{CV}}}$$

where

$$F_{\text{ext}} = F_p + (W_L + W_{\text{belt}})(\sin\beta + \mu\cos\beta)$$

3.4   Torque Required for the Motion

If we look at the drive-train from the motor side, we can see that there are two torques involved as shown in Figure 3.21. The motor applies a torque (T_m) which works against the load torque reflected to the motor shaft $(T_{\text{load}\rightarrow\text{M}})$. We can apply Newton's second law as

$$\sum T = J_{\text{total}}\ddot{\theta}_m$$

Torque balance on the motor shaft gives

$$T_m - T_{\text{load}\rightarrow\text{M}} = J_{\text{total}}\ddot{\theta}_m \tag{3.34}$$

Or,

$$T_m = J_{\text{total}}\ddot{\theta}_m + T_{\text{load}\rightarrow\text{M}} \tag{3.35}$$

where T_m is the torque needed from the motor to achieve the motion; J_{total} is the inertia of all transmission elements + motor + load *reflected to the motor shaft*; $\ddot{\theta}_m$ is the motor shaft angular acceleration; and $T_{\text{load}\rightarrow\text{M}}$ is the torque demand from the motor due to all external loads *reflected to the motor shaft*.

The load on the motor is from the external torques due to the following

- Friction (T_f),
- Gravity (T_g), and
- Process torques (T_{process}) (e.g., acting on the tool of the machine during assembly).

$$T_{\text{ext}} = T_f + T_g + T_{\text{process}} \tag{3.36}$$

When the load is directly connected to the motor

$$T_{\text{load}\rightarrow M} = T_{\text{ext}}$$

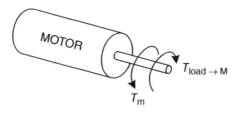

Figure 3.21   Torques on the motor shaft

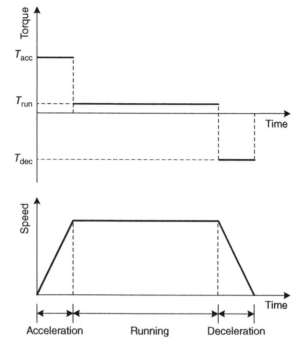

Figure 3.22   Phases of the motion profile and torque in each phase

Otherwise, T_{ext} must be reflected to the motor shaft through the gearbox and/or transmission mechanism to calculate $T_{\text{load}\to\text{M}}$. The motor torque T_m needed to achieve the motion profile depends on the phase of the motion (Figure 3.22) [23].

3.4.1   Acceleration (Peak) Torque

In the acceleration phase, Equation (3.35) can be rewritten as follows

$$T_{\text{acc}} = J_{\text{total}}\, \ddot{\theta}_m + T_{\text{load}\to\text{M}} \tag{3.37}$$

As shown in Figure 3.22, while accelerating the load, the motor tends to use the highest amount of torque since it works against all loads and accelerates all inertia in the system. Therefore, the acceleration torque is often called the *peak torque* and denoted by T_{peak}.

3.4.2   Running Torque

Once the load is traveling at a constant running speed, the acceleration is equal to zero and the motor is not working against the system inertia (Figure 3.22). Therefore,

$$T_m - T_{\text{load}\to\text{M}} = 0$$

Then, the torque needed from the motor becomes

$$T_{\text{run}} = T_{\text{load} \to \text{M}} \tag{3.38}$$

3.4.3 Deceleration Torque

In the deceleration phase of the motion, the acceleration is negative, hence the torque is negative as seen in Figure 3.22. Therefore, from Equation (3.34)

$$T_{\text{dec}} - T_{\text{load} \to \text{M}} = -J_{\text{total}}\, \ddot{\theta}_m$$

or

$$T_{\text{dec}} = T_{\text{load} \to \text{M}} - J_{\text{total}}\, \ddot{\theta}_m \tag{3.39}$$

3.4.4 Continuous (RMS) Torque

Since the torque requirement varies depending on the phase of the motion, an *average* continuous torque value can be computed by finding the root-mean-square (RMS) of all the torques required in one cycle of the motion. In a more generalized motion profile, the load may have some rest periods (called *dwell*) in its cycle as shown in Figure 3.23. Then,

$$T_{\text{RMS}} = \sqrt{\frac{T_{\text{acc}}^2 \cdot t_a + T_{\text{run}}^2 \cdot t_m + T_{\text{dec}}^2 \cdot t_d + T_{\text{dw}}^2 \cdot t_{\text{dw}}}{t_a + t_m + t_d + t_{\text{dw}}}} \tag{3.40}$$

where T_{acc}, T_{run}, T_{dec}, and T_{dw} are the torques required in the acceleration, running, deceleration, and dwell phases of the motion, respectively. Similarly, t_a, t_m, t_d and t_{dw} are the duration of each phase. The dwell torque may be zero if the axis is stopped and is not working against any external force during the dwell. It may also be nonzero such as in case of an axis with vertical load where torque must be applied to hold the load even though the axis is stopped.

■ **EXAMPLE 3.4.1**

The gantry machine in Figure 3.24 carries a 20 kg load. The X-axis of the machine uses two 404XE-T13-VL linear tracks in parallel by Parker Hannifin Corp [30]. The Y-axis and Z-axis

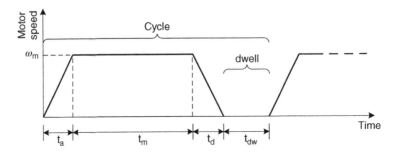

Figure 3.23 Cyclical motion profile with dwell period

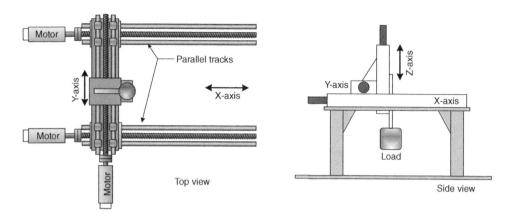

Figure 3.24   Gantry machine with two parallel linear tracks for X-axis

use 403XE-T04 and 402XE-T04 linear tracks by the same manufacturer [31]. The X- and Y-axis tracks have 10 mm screw leads. The Z-axis has 5 mm screw lead. Each axis is equipped with a BE230D motor from the same manufacturer [27]. Each motor is connected to its axis using a cylindrical aluminum coupler with 1.125 in outer diameter and 1.5 in length. Referring to the data sheets, calculate the acceleration and continuous torque required to move each linear track of the X-axis at 250 cts/ms speed for 1 s. Acceleration time is $t_a = 50$ ms.

Solution

1. Find maximum angular acceleration $\ddot{\theta}_{motor}$ of the motor

$$\omega_m = 250 \frac{\text{cts}}{\text{ms}} \cdot \left(\frac{1 \text{ rev}}{8000 \text{ cts}} \right) \left(\frac{1000 \text{ ms}}{1 \text{ s}} \right) \left(\frac{2\pi \text{ rad}}{1 \text{ rev}} \right)$$
$$= 196.35 \text{ rad/s}$$

Angular acceleration of the motor is

$$\ddot{\theta}_{motor} = \alpha_m = \frac{\omega_m}{t_a} = \frac{196.35}{0.05} = 3927 \text{ rad/s}^2$$

2. The total inertia seen by the motor is

$$J_{total} = J_{motor} + J_{coupling} + J_{ref}^{trans}$$

(a) Motor

Referring to the motor datasheet for BE230D, the motor inertia is

$$J_{motor} = 5.2 \times 10^{-6} \text{ kg-m}^2$$

(b) Coupling

The coupling connects the motor shaft to the ball-screw in the linear track. It is made out of aluminum ($\rho = 2810 \text{ kg/m}^3$). From the motor datasheet, the motor shaft diameter is found as 0.375 in. This is also the inner diameter of the coupler. Then, we can calculate the coupler inertia as a hollow cylinder as

$$J_{coupling} = \frac{\pi L \rho}{2}(r_o^4 - r_i^4)$$

$$= \frac{\pi \cdot (1.5 \cdot 0.0254) \cdot 2810}{2}((0.0254 \cdot 1.125/2)^4$$

$$- (0.0254 \cdot 0.375/2)^4)$$

$$= 6.92 \times 10^{-6} \, \text{kg-m}^2$$

(c) Inertia reflected by ball-screw transmission consists of the ball-screw inertia and the load inertia reflected to the input shaft of the track. From Equation (3.26)

$$J_{ref}^{trans} = J_{screw} + J_{L \rightarrow in}$$

The datasheet for the 404XE-T13-VL lists the ball-screw diameter as 16 mm. The T13-VL is 533 mm long and has 90% efficiency and 0.1 friction coefficient. For this example, the ball-screw is approximated as a solid round shaft with an estimated pitch diameter of 13 mm and length of 533 mm. The material density is 7800 kg/m^3 (steel). Then, the inertia of the screw as a solid cylinder is

$$J_{screw} = \frac{\pi \cdot L \cdot \rho \cdot r^4}{2}$$

$$= \frac{\pi \cdot 0.535 \cdot 7800 \cdot (13/2000)^4}{2}$$

$$= 1.17 \times 10^{-5} \, \text{kg-m}^2$$

The total weight carried by each track of the X-axis consists of the weight of the Y-axis, Z-axis, and the load. Since the X-axis is made out of two parallel tracks, the total weight will be distributed between them. When the load and the Z-axis are in the middle, the total weight is divided equally between the two tracks of the X-axis. But if the Z-axis and the load are moved all the way to one end of the Y-axis, the X-axis track close to that end will have to carry almost the entire weight of the load and the Z-axis. For the worst case scenario torque calculations, we will take the total weight as the load carried by each track of the X-axis. Then, W_L is the total weight given by

$$W_L = W_{Y-axis} + W_{Z-axis} + W_{load}$$

$$= (3.55 + 1.81 + 2 \cdot 0.67 + 20) \cdot 9.81$$

$$= 261.93 \, \text{N}$$

where 0.67 kg is the mass of a motor, 3.55 kg and 1.81 kg are the masses of the Y-axis and Z-axis from the data sheets, respectively. The weight of the carriage W_C for the

X-axis is $0.495 \cdot 9.81 = 4.856\,\text{N}$.

$$J_{L \to in} = \left(\frac{1}{\eta N_s^2}\right)\frac{W_L + W_C}{g}$$

$$= \left(\frac{1}{0.9 \cdot (2\pi \cdot 100)^2}\right)\frac{261.93 + 4.856}{9.81}$$

$$= 7.654 \times 10^{-5}\,\text{kg-m}^2$$

where $N_s = 2\pi p$ is the transmission ratio for the ball screw. The X-axis has 10 mm/rev lead, which is 0.010 m/rev. Therefore, $p = 1/0.010 = 100\,\text{rev/m}$. Hence,

$$J_{ref}^{trans} = J_{screw} + J_{L \to in}$$

$$= 1.17 \times 10^{-5} + 7.654 \times 10^{-5}$$

$$= 8.824 \times 10^{-5}\,\text{kg-m}^2$$

Then, the total inertia is

$$J_{total} = J_{motor} + J_{coupling} + J_{ref}^{trans}$$

$$= 5.2 \times 10^{-6} + 6.92 \times 10^{-6} + 8.824 \times 10^{-5}$$

$$= 1.004 \times 10^{-4}\,\text{kg-m}^2$$

3. Load torque reflected by the ball-screw to the motor consists of friction force, gravitational loading, and process forces. From Equation (3.28), the friction force is

$$F_f = \mu \cdot (W_L + W_C)\cos(\beta)$$

$$= 0.01 \cdot (261.93 + 4.856)$$

$$= 2.67\,\text{N}$$

where β is the angle of the track with the horizontal, which is zero. Similarly, the gravitational loading that the track has to work against can be found from

$$F_g = (W_L + W_C)\sin(\beta)$$

$$= 0\,\text{N}$$

In this example, the system is not pushing against any external forces. Therefore, $F_p = 0$. Then, from Equation (3.27), the total external force is

$$F_{ext} = F_f + F_g + F_p$$

$$= 2.67\,\text{N}$$

The load torque reflected to the motor by the ball screw is given by

$$T_{load \to in} = \frac{F_{ext}}{\eta N_s}$$

$$= \frac{2.67}{0.9 \cdot 2\pi \cdot 100}$$

$$= 0.0047 \, \text{Nm}$$

4. The acceleration (peak) torque can be found from

$$T_{acc} = J_{total} \cdot \ddot{\theta}_{motor} + T_{load \to in}$$

$$= 1.004 \times 10^{-4} \cdot 3927 + 0.0047$$

$$= 0.3988 \, \text{Nm}$$

The running torque is simply equal to the load torque reflected to the motor

$$T_{run} = T_{load \to in} = 0.0047 \, \text{Nm}$$

Deceleration torque is

$$T_{dec} = T_{load \to in} - J_{total} \cdot \ddot{\theta}_{motor}$$

$$= 0.0047 - 1.004 \times 10^{-4} \cdot 3927$$

$$= -0.3894 \, \text{Nm}$$

The continuous torque (RMS) is calculated using Equation (3.40) as

$$T_{RMS} = 0.119 \, \text{Nm}$$

where $t_a = t_d = 0.050 \, \text{s}$, $T_{dw} = 0 \, \text{N}$, $t_{dw} = 0 \, \text{s}$, and $t_m = 1 \, \text{s}$.

3.5   Motor Torque–Speed Curves

Most industrial motion control systems use three-phase AC servomotors and/or vector-duty AC induction motors. Table 3.2 provides a comparison of typical stocked three-phase (3ϕ) motors used in industrial motion control with AC drives. It should be noted that there is a significant overlap between the motor types in terms of characteristics and and power ranges. Furthermore, there are small AC servomotors available down to the range of 0.01 HP (10 W).

Motor catalogs provide torque–speed curves for each motor along with tables with extensive electrical and mechanical data. The torque–speed curves and the data are provided for a particular motor voltage, ambient temperature for the application, and the motor+drive combination. In selecting a motor, one must pay attention to these conditions and how well they match the requirements of the particular application in hand.

Table 3.2 Comparison of vector-duty AC induction motors and AC servomotors operated with 460 VAC AC drives

	Vector-duty Three-phase induction motor	Three-phase AC servomotor
Typical Power @460VAC (HP)	0.25–30	1–16
Cont. torque (lb in)	9–12 000	9–257
Max speed (rpm)	2× base speed	5000
Min speed (with full load)	0	0
Motor inertia	Medium	Low
Motor size	Medium	Small
Feedback	Encoder	Encoder
Speed regulation (with full load)	1000:1	10 000:1
Features	Full torque at zero speed; slower acceleration; typical applications include print cylinders, winders, positioning conveyors	Smooth motion; rapid acceleration; most popular in motion control; excellent dynamic performance; typical applications include pick-and-place, indexing, flying shears, high speed material handling

3.5.1 Torque–Speed Curves for AC Servomotors

Figure 3.25 shows a typical torque–speed curve for an AC servo motor under closed-loop control of an electric drive. There are two regions: (i) continuous operation, and (ii) intermittent operation.

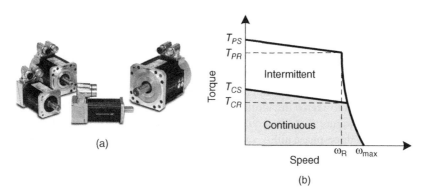

(a)

(b)

Figure 3.25 Generic torque–speed curves for an AC servo motor under closed-loop control of an electric drive. (a) AC servo motors (Reproduced by permission of Emerson Industrial Automation) [12]. (b) Torque-speed curves for a generic motor

The *continuous operation region* indicates the torque available in various speeds for safe operation of the motor for long periods of time. The *intermittent operation region* shows much higher torque levels that the motor can deliver but only for very brief periods of time. For example, the intermittent operation can be for up to about 30 s during the acceleration phase of the motion where peak torque is required. If the motor were operated at such high torque levels for longer periods of time, then the windings would overheat resulting in permanent damage to the motor.

The maximum speed ω_{max} is the speed of the motor at full supply voltage and with no load. For a servo motor, this can be 5000–6000 rpm. The peak stall torque T_{PS} is the maximum torque the motor can provide *for a brief period of time* when it is stalled. The continuous stall torque T_{CS} is the maximum torque the motor can provide for longer periods of time at stall. Most motor datasheets list a rated speed ω_R (at the corner of the top curve), rated peak torque T_{PR}, and rated continuous torque T_{CR}. The *rated* values are simply convenient reference points where the manufacturer is providing data. To verify that the rated speed is reached, the manufacturer tests the motor using a rated voltage and operates it with a rated torque. Almost all motor datasheets provide the torque–speed curves for the motors. If these curves are available, they should be used in the motor selection process. However, some datasheets provide the motor data only as a table. In this case, they can be approximately sketched as shown in Figure 3.25, by using the data given in these tables.

3.5.2   Torque–Speed Curves for AC Induction Motors

Three-phase AC induction (squirrel cage) motors are extensively used in industry. A standard induction motor is designed to operate at constant speed when directly connected to three-phase utility supply across-the-line or direct-on-line (DOL) operation. In motion control applications, speed and torque are variable. In recent years, advances in microprocessor technology, power electronics devices, and modern control schemes, such as field-oriented control (or vector control), made it possible to use induction motors in variable speed applications.

A standard induction motor is totally enclosed with a fan mounted on the back of it. As the motor runs at its constant rated speed, the fan generates sufficient airflow across the fins on the outer case of the motor to prevent overheating. If such a motor is operated in low speeds, the motor will overheat. Therefore, a special type of induction motor called *inverter-duty* or *vector-duty* must be used in variable speed applications. These motors have special insulation in their windings and have been specifically designed to operate with the power electronics in the drive. Some of these motors come with an encoder, which enables using them in closed-loop control with a vector control drive.

Data for a vector-duty motor are supplied as a table of values in manufacturer catalogs [21]. The data can be used to *sketch* the torque–speed curves for an *induction motor under vector drive control* as in Figure 3.26. It should be noted that if the motor is operated by directly connecting it to the three-phase utility supply (DOL), the torque–speed curve will be different as explained in Section 4.4.4 of Chapter 4.

The *full-load speed* is the speed of the motor under full load condition. This speed is obtained when the input voltage is applied at 60 Hz, which is called the *base frequency*. Recall that a typical induction motor is designed to operate direct-on-line with 60 Hz utility supply. The *full-load torque* is the torque the motor develops to produce its rated horsepower at full-load speed. The *breakdown torque* is the maximum torque the motor will develop at the rated

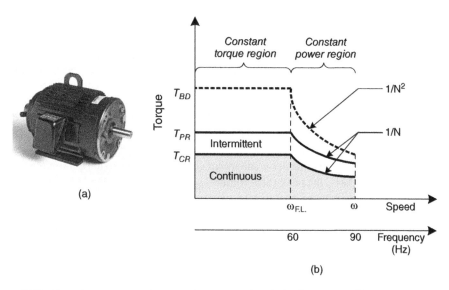

Figure 3.26 Generic torque–speed curves for a vector-duty induction motor with vector-control drive. (a) Vector-duty AC induction motor (Reproduced by permission of Marathon™ Motors, A Regal Brand) [21]. (b) Torque-speed curves for a *generic* motor

voltage and speed while rotating. If for some reason the load torque exceeds the breakdown torque, the motor will stall [19].

An induction motor with vector control drive can produce constant torque at any speed from zero up to the rated speed. Therefore, this part of the torque–speed curve is called *constant torque region*. The drive achieves this by adjusting the input voltage and frequency so that their ratio (V/f) remains constant. The rated voltage of the motor is applied at the base frequency. If the input frequency is increased beyond the base frequency (60 Hz), the voltage remains constant. As a result, the V/f reduces inversely with the frequency. The region beyond the base frequency is called *constant power region* or *field weakening region*. The power generated by the motor is given by

$$\mathbb{P} = T_{FL} \cdot \omega_{FL}$$

where T_{FL} and ω_{FL} are the full-load torque and speed of the motor, respectively. At a speed ω beyond the rated speed, the torque T must drop so that the power remains constant

$$\mathbb{P} = T_{FL} \cdot \omega_{FL} = T \cdot \omega$$

Then, torque T at a speed ω beyond the base speed can be found as

$$T = (T_{FL}\omega_{FL}) \cdot \frac{1}{\omega}$$

Since the full-load speed corresponds to the base frequency N_b, we can rewrite the same equation as

$$T = (T_{FL}N_b) \cdot \frac{1}{N} \tag{3.41}$$

where N is a frequency beyond the base frequency. Most induction motors can deliver up to 1.5 times the base speed (90 Hz) in the constant power region [21]. Some motors can exceed this limit to deliver up to two times the base speed. Factors such as bearings, lubrication, rotor balancing, frame construction, and load connection impose maximum safe mechanical speed limits on the motor.

The drive is designed to impose current limits on the motor to protect the motor and the power electronic devices in the drive. As a result, the drive and motor combination will determine the continuous and peak torque available for a motion application. In Figure 3.26, the continuous rated torque T_{CR} can be taken as the lesser of the full-load torque of the motor or the continuous torque generated by the drive and motor combination due to the continuous current limit of the drive. Similarly, the peak rated torque T_{PR} can be taken as the lesser of the breakdown torque of the motor or the peak torque generated by the drive and motor combination due to the peak current limit of the drive. The drive should not exceed the breakdown torque of the motor since the motor will stall. Therefore, the breakdown torque creates an absolute maximum limit a motor can deliver for short periods of time without overheating. Furthermore, beyond the base frequency, the breakdown torque T_{BD} falls proportional to $1/N^2$, where N is the applied frequency.

◼ EXAMPLE 3.5.1

A 5 HP Black Max® vector-duty motor with 460 VAC supply voltage at 60 Hz base frequency has 1765 rpm F.L. speed, 7A F.L. current, 14.9 lb ft F.L. torque and 70 lb ft breakdown (B.D.) torque [21]. Sketch the torque–speed curves for this drive and motor combination if the motor is connected to a drive with 5A continuous and 10A peak current.

Solution
The motor will draw 7 A current when providing 14.9 lb ft of F.L. torque. Since the drive can only supply 5 A continuous current, the F.L. torque needs to be de-rated. We can find the linear scaling factor from

$$K_t = \frac{14.9}{7} = 2.13 \,(\text{lb ft})/\text{A}$$

With the 5 A drive continuous current, the motor can provide $2.13 \cdot 5 = 10.7$ lb ft torque. Since this value is less than the 14.9 lb ft F.L. torque, $T_{CR} = 10.7$ lb ft. If the drive had more continuous current than the 7 A F.L. current of the motor, the drive current would need to be limited in its software to the F.L. current of the motor to prevent any damage to the motor due to overheating under continuous operation conditions. Then, the T_{CR} would have been equal to the rated F.L. torque of the motor.

Using Equation (3.41), we can find the continuous torque at $N = 90$ Hz as

$$T = 10.7 \cdot 60/90$$

$$= 7.1 \,\text{lb ft}$$

Similarly, at the peak current of the drive, the motor will develop $2.13 \cdot 10 = 21.3$ lb ft of peak torque. Since this value is less than the 70 lb ft BD torque, $T_{PR} = 21.3$ lb ft. If the drive had more peak current resulting in more torque than the BD torque of the motor, the drive current would need to be limited in software since the motor cannot be operated beyond its breakdown torque due to stall conditions. Then, the T_{PR} would have been equal to the BD torque of the motor.

Using Equation (3.41), we can find the peak torque at $N = 90\,\text{Hz}$ as $14.2\,\text{lb ft}$. Figure 3.27 shows the torque–speed curves for this drive and motor combination.

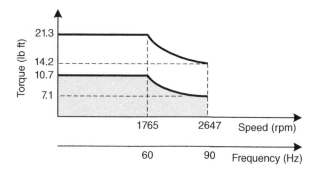

Figure 3.27   Torque–speed curve for a 5 HP motor and drive combination

3.6   Motor Sizing Process

Motor sizing is the process of selecting the best motor for a motion control application. The motion profile and system inertia help determine the motor speed, acceleration and the required torques as explained in Section 3.4. Other factors such as cost, physical size of the motor, and drive power requirements must also be considered.

Four primary factors are considered in motor sizing:

1. Inertia ratio (J_R)
2. Motor speed (ω_m)
3. Peak torque (T_{peak}) at the motor speed
4. RMS torque (T_{RMS}) at the motor speed.

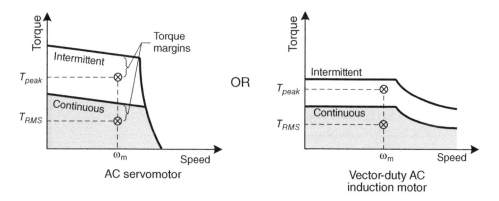

Figure 3.28   Both RMS and peak torque requirements of the motion must be within the capabilities of the motor at the desired motor speed for proper operation

The design task is to find the smallest motor that can produce the speed and torques required for the motion. Given a desired motion profile for the axis, the sizing process starts by finding the motor speed needed to achieve the running speed of the axis. Then, the peak and RMS torques for the motion need to be calculated. Next, a motor must be selected such that both the peak and RMS torques are within the peak and continuous torque capabilities of the motor at the desired motor speed as shown in Figure 3.28. Finally, we must also ensure that the inertia ratio criterion is met by the selected motor. Engineering data for a motor are provided in its catalog including the motor inertia (J_m), rated peak and continuous torques and rated speed. Often, the catalogs also provide the torque–speed curves.

3.7 Motor Selection for Direct Drive

In direct drive systems, the motor is directly coupled to the load as shown in Figure 3.29. Motor selection involves finding a motor that can deliver the peak torque (T_{peak}) and continuous torque (T_{RMS}) required for the load to follow the motion profile.

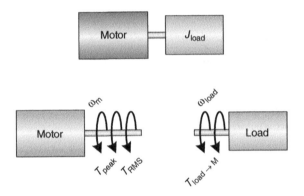

Figure 3.29   Direct drive where motor is directly coupled to the load

Section 3.4 explained how to calculate the peak and continuous (RMS) torques required for a particular motion profile for the load. Using the T_{peak}, T_{RMS}, and the running speed $\dot{\theta}_m$ (or ω_m) of the motor from the velocity profile, we can pick a motor that can deliver these torques at the running speed.

MOTOR SELECTION PROCEDURE FOR DIRECT DRIVE

1. Find the running speed ω_m of the motor from the desired motion profile for the load. Calculate T_{peak} and T_{RMS} that the motor should provide. At this point the motor is unknown. Therefore, use $J_m = 0$ in the torque calculations for now.

MOTOR SELECTION PROCEDURE FOR DIRECT DRIVE (CONTINUED)

2. From a motor catalog pick the smallest possible motor that can provide these torques at the running speed ω_m of the motor. Motor data may be provided as torque–speed curves or in tabular format. Accept the motor selection if at ω_m

$$T_{\text{peak}} \leq T_{\text{PR}} \text{ and } T_{\text{RMS}} \leq T_{\text{CR}}$$

An electric drive will be used to run the motor. Therefore, a suitable drive must be selected that can provide the currents necessary to run the motor at the peak and continuous torques. If the drive has less current than the torque levels require, the motor T_{PR} and T_{CR} should be de-rated (scaled down) linearly since the drive will limit the current, hence the motor torques.

3. Recalculate T_{peak} and T_{RMS} with the newly selected motor's inertia J_m.

4. Check the selection criterion in Step 2 to make sure that the motor can still deliver the expected performance with some margin. In practice 30% margin for T_{RMS} and 50% margin for T_{peak} are common. It is desirable to have some margin (extra torque capability) for adjustments, if conditions change during the commissioning of the machine.

5. Calculate the inertia ratio J_R and make sure that it meets the desired criterion. Typically, $J_R \leq 5$ is used but this also depends on the desired system performance. If the criterion is not met, select a motor with larger inertia (bigger motor) or reduce the load inertia and repeat steps 3–5.

3.8 Motor and Transmission Selection

This axis configuration is shown in Figure 3.7. The load is connected to the motor via a transmission mechanism. The mechanism reflects the load torque, inertia, and motion to its input shaft (motor) as discussed in Section 3.3.

The required peak and RMS torques can be calculated as explained in Section 3.4. In these calculations, the J_{total} term can be found from Equation (3.21), where the reflected inertia $J_{\text{ref}}^{\text{trans}}$ is uniquely defined for each transmission type in Section 3.3. Furthermore, the load reflected to the input shaft of the mechanism ($T_{\text{load} \rightarrow \text{in}}$) is the same as the load reflected to the motor shaft ($T_{\text{load} \rightarrow \text{M}}$) in the equations of Section 3.4 since the motor is connected to the input shaft of the transmission.

■ **EXAMPLE 3.8.1**

The flying shear machine shown in Figure 2.15.a in Chapter 2 will be redesigned to use a belt-drive actuator instead of the lead screw. The designer plans to use an ERV5 rodless actuator (belt drive) [26] to replace the lead screw actuator. The cutting tool assembly weighs 35 lb. The axis has a travel length of 48 in. The cutter needs to travel at $V_{\text{fwd}} = 16\,\text{in/s}$ speed in synchronization with the wood board while cutting the material. The trapezoidal motion profile in Figure 2.15b has $t_1 = 500\,\text{ms}$ and $t_2 = 2.5\,\text{s}$. The return motion is completed in $t_3 = 2\,\text{s}$. Select an AC servomotor for the axis from the BE Series servo motor catalog by Parker Hannifin Corp. [27].

Solution

1. The motor will be directly coupled to the belt drive. Therefore, the running speed of the motor in the forward motion can be found by dividing the linear speed of the axis V_{fwd} by the radius of the pulley inside the rodless actuator. From the catalog [26] $r_{ip} = 0.6265\,\text{in}$. Then, $\omega_{m.fwd} = 16/0.6265 = 25.54\,\text{rad/s}$.

   The reflected inertia can be found from Equation (3.32). The catalog specifies the total rotational inertia for the base unit of the belt drive as reflected to its input shaft. It also provides the inertia due to each additional 100 mm of stroke length (mainly due to the added belt length). To take advantage of these data, we can rearrange Equation (3.32) as follows:

$$J_{\text{ref}}^{\text{trans}} = J_{\text{IP}} + J_{\text{carriage}\rightarrow\text{in}} + J_{\text{belt}\rightarrow\text{in}} + J_{\text{LP}} + J_{\text{load}\rightarrow\text{in}}$$

$$= \left[2J_P + \frac{1}{\eta N_{\text{BD}}^2} \left(\frac{W_C + W_{\text{belt}}}{g} \right) \right] + \frac{1}{\eta N_{\text{BD}}^2} \left(\frac{W_L}{g} \right)$$

$$= J_{\text{BeltDrive}\rightarrow\text{in}} + \frac{1}{\eta N_{\text{BD}}^2} \left(\frac{W_L}{g} \right)$$

where $J_{\text{BeltDrive}\rightarrow\text{in}}$ is the total reflected rotational inertia of the base unit for the belt drive. The base unit inertia for the ERV5 is 20.71 oz-in². For each additional 100 mm of travel, the additional inertia is specified as 0.03 oz-in². Then, for the axis with 48 in (1219 mm) of travel length the total reflected inertia is $J_{\text{BeltDrive}\rightarrow\text{in}} = 20.71 + 0.03 \cdot 12.19 = 21.076\,\text{oz-in}^2$. The reflected inertia by the belt drive with the load is as follows:

$$J_{\text{ref}}^{\text{trans}} = J_{\text{BeltDrive}\rightarrow\text{in}} + \frac{1}{\eta N_{\text{BD}}^2} \left(\frac{W_L}{g} \right)$$

$$= \frac{21.076}{386} + \frac{1}{0.9 \cdot 1.596^2} \left(\frac{35 \cdot 16}{386} \right)$$

$$= 0.6873\,\text{oz-in-s}^2$$

   Note that the first term was divided by gravity g for unit conversion.

   The axis accelerates to V_{fwd} speed in $t_1 = 500\,\text{ms}$. Therefore, the angular acceleration of the motor in the forward motion should be $\ddot{\theta}_{m.fwd} = 25.54/0.5 = 51.08\,\text{rad/s}^2$.

   The load torque reflected to the input shaft of the belt drive is given by Equation (3.33) where F_{ext} can be calculated from Equation (3.29). The belt drive is horizontal ($\beta = 0$).

There is no external process force ($F_p = 0$) and the carriage weight is specified in the catalog as 2.99 lb. Then, with a friction coefficient of $\mu = 0.1$ the external force is $F_{ext} = (35 + 2.99) \cdot 0.1 \cdot 16 = 60.78$ oz. From Equation (3.33), $T_{load \to in} = 42.31$ oz in.

Since the motor is unknown, $J_m = 0$. Then, from Equation (3.37) we can calculate the peak torque in the forward motion as $T_{peak.fwd} = 4.85$ lb in. We also need to calculate the peak torque required to accelerate the cutting tool assembly in the reverse direction to return it to the starting position for the next cut. We can use the same equation but we must first find the angular acceleration of the motor in the return motion, which is equal to

$$\ddot{\theta}_{m.ret} = \frac{\omega_{m.ret}}{t_3/2}$$

where $\omega_{m.ret} = V_{ret}/r_{ip}$ is the maximum motor speed in the return motion and V_{ret} is the return velocity in Figure 2.15b. Since the cutting tool must return to the same starting position after it makes a cut, the distance it travels in the forward motion should be equal to the return distance. Therefore, by equating the area under the trapezoidal velocity profile to the area under the triangular return velocity profile, we can find the return velocity as

$$V_{ret} = 2V_{fwd}\frac{t_1 + t_2}{t_3}$$

Substituting the values in this equation gives $V_{ret} = 48$ in/s and $\ddot{\theta}_{m.ret} = 76.62$ rad/s^2. Then, from Equation (3.37) the peak torque for the return motion can be calculated as $T_{peak.ret} = 5.94$ lb in. Note that it requires more peak torque to quickly accelerate and return the cutter assembly to its start position in shorter time.

Refering to Figure 2.15b and Equation (3.40), the RMS torque for the forward, return and the entire motion can be calculated using

$$T_{RMS.fwd} = \sqrt{\frac{T^2_{peak.fwd} \cdot t_1 + T^2_{run} \cdot t_2 + T^2_{dec.fwd} \cdot t_1}{t_1 + t_2 + t_1}}$$

$$T_{RMS.ret} = \sqrt{\frac{T^2_{peak.ret} \cdot t_3/2 + T^2_{dec.ret} \cdot t_3/2}{t_3}}$$

$$T_{RMS} = \sqrt{\frac{T^2_{RMS.fwd} \cdot (t_1 + t_2 + t_1) + T^2_{RMS.ret} \cdot t_3}{t_1 + t_2 + t_1 + t_3}}$$

where T_{run} is found from Equation (3.38), $T_{dec.fwd}$ and $T_{dec.ret}$ are found from Equation (3.39). Substituting all values gives $T_{RMS} = 3.44$ lb in for the entire motion.

2. From the BE Series servo motor catalog [27], we need to identify the smallest motor that can provide these torques. The NEMA size 23 BE232F motor has $T_{PR} = 28.9$ lb in and $T_{CR} = 8.5$ lb in. Since both of these values are greater than the calculated torques, this motor is selected.

3. The inertia of the selected motor is $J_m = 1.5 \times 10^{-4}$ lb-in-s^2. This value must be multiplied by 16 to convert it into oz-in-s^2 and then used in Equations (3.37), (3.38) and (3.39) to

recalculate the peak and RMS torques. The RMS torque for the entire motion becomes $T_{RMS} = 3.44$ lb in. The largest peak torque is during the return phase and is found as $T_{peak} = 5.95$ lb in.

4. The recalculated values are still less than the peak and continues torques of the selected motor. Therefore, the motor is kept. The torque margins are calculated from

$$\text{Peak Torque Margin} = \frac{T^m_{peak} - T^{app}_{peak.ret}}{T^m_{peak}}$$

$$= \frac{28.9 - 5.95}{28.9}$$

$$= 79\%$$

$$\text{RMS Torque Margin} = \frac{T^m_{RMS} - T^{app}_{RMS}}{T^m_{RMS}}$$

$$= \frac{8.5 - 3.44}{8.5}$$

$$= 59\%$$

Both the peak and RMS torque margins are more than the practical 50% and 30%, respectively. Therefore, the selected motor can easily accomplish the desired motion.

3.9   Gearboxes

A gearbox is coupled to a motor to reduce its speed and to multiply its torque. As a result, smaller motors can move larger loads. A gearbox also allows the machine designer to adjust the inertia ratio. The inertia ratio compares the total load inertia to the inertia of the motor. If a gearbox is used in the design, the reflected load inertia is reduced by the square of the gearbox ratio (Equation (3.15)). Often it is less expensive to use a motor and gearbox (gearhead) than to use a larger motor in direct-drive configuration [25]. Use of a gearbox in the design can also increase the torsional stiffness of the drive-train.

Most common gearboxes used in industrial motion control applications are (1) Planetary servo gearboxes (gearheads), and (2) Worm gear speed reducers.

3.9.1   Planetary Servo Gearheads

Planetary gearheads are primarily used with servo motors. They are available in standard NEMA and IEC sizes for easy mounting on servo motors [8, 16, 29, 37]. The planetary gearheads offer low backlash, higher output torques and smaller sizes but cost more. There are *in-line* and *right-angle* models with gear ratios of 3:1 to 100:1 (Figure 3.30). Typically, ratios up to 10:1 are available as single-stage gearheads. Higher ratios require additional planetary stages in the unit, which increases the cost.

The following primary factors need to be consider in selecting a gearhead:

1. Mode of operation (continuous or cyclical)
2. Gear ratio (N_{GB})

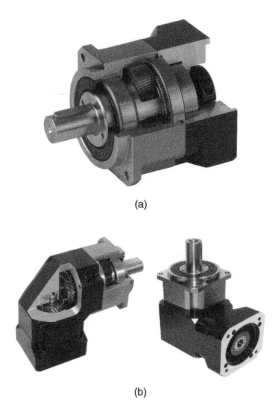

(a)

(b)

Figure 3.30   Servo gearheads (Reproduced by permission of Apex Dynamics USA). (a) Inline gearhead [3]. (b) Right-angle gearhead [4]

3. Nominal and maximum acceleration output torques (T_{2N}, T_{2B})
4. Nominal and maximum input speeds (ω_{1N}, ω_{1B})
5. Mechanical compatibility with the motor (mounting flange, shaft diameter, bolt patterns, etc.).

In addition, most gearhead catalogs specify an *emergency stop (E-stop) output torque*. The motion controller will bring a motor to a controlled stop with rapid deceleration when an E-stop condition occurs. Depending on the load and the axis speed, the E-stop can require significantly high torques to stop the load in a very short period of time. The typical E-stop output torque level is about 3–4 times the nominal output torque level for a gearhead. But each unit can allow only a limited number of such high loads in its lifetime (such as maximum of 1000 times).

3.9.2   Worm Gear Speed Reducers

Worm gear speed reducers are primarily used with AC induction motors. They are available in standard NEMA and IEC sizes for easy mounting on the motors [6, 8, 10].

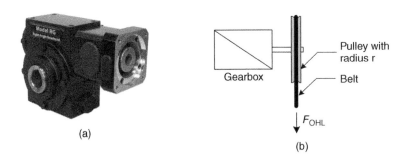

(a)

(b)

Figure 3.31 Worm gear speed reducer. (a) Right-angle worm gear reducer for AC induction motors (Reproduced by permission of Cone Drive Operations, Inc.) [9]. (b) Overhung load

Gearboxes with worm gear come in standard and low backlash models. They can tolerate high shock loads but are less efficient than other forms of gears (60–95% efficiency). They are available in right-angle configuration (Figure 3.31a) with 5:1–60:1 gear ratio in single stage.

The following primary factors need to be consider in selecting a worm gear speed reducer:

1. Gear ratio (N_{GB})
2. Case size (such as 15, 20, 25, 30, etc.)
3. Level of precision (low or standard backlash)
4. Nominal input speed ($\omega_{GB.nom}$)
5. Input horsepower or output torque
6. Overhung load (F_{OHL})
7. Service factors (S_f)
8. Mechanical compatibility with the motor (mounting flange, shaft diameter, bolt patterns, output shaft options, etc.)

The required torque for the application comes from analyzing the motion profile and the drive-train. *Service factors* take into account various operational factors such as cyclical operation, load characteristics and thermal conditions. Since such conditions adversely affect the performance and life of the gearbox, the service factors are used to either *increase* the required torques for the application by multiplying the calculated values or to *de-rate* the horsepower/torque data given in the catalog for a particular gearbox by dividing them by the service factors. These adjusted values are sometimes called "design" horsepower/torque since the adjusted values are used in the rest of the gearbox selection process.

An important consideration is how the gearbox is mounted to the motor and to the driven load. Options include sprockets, pulleys, flexible couplings and hollow bore output shaft for direct-drive. When the load is connected with a sprocket or pulley, a radial force is applied to the shaft of the gearbox by the pull of the chain or belt as shown in Figure 3.31b. This is a bending force often called *overhung load* F_{OHL}

$$F_{OHL} = \frac{T \cdot K_{LF}}{r} \tag{3.42}$$

where T is the torque on the shaft, r is the radius of the pulley/sprocket and K_{LF} is the overhung load factor, which is specified in the gearbox catalog for the loading condition.

Misalignment of the reducer shaft with the load shaft can also cause overhung loading even without any pulley or sprocket. The datasheets specify the maximum allowable overhung load for each gearbox family. The overhung load can be reduced by increasing the diameter of the pulley/sprocket, by mounting the pulley closer to the case of the reducer and by using flexible couplings to minimize misalignment loading. Just as in the servo gearheads, the worm gear speed reducers also have higher E-stop torque levels and allow only a limited number of such overloads in the lifetime of a gearbox.

Gearbox catalogs provide horsepower (HP) ratings. Input HP can be calculated from

$$HP_{in} = \frac{T_{in}\omega_w}{63\,000} \tag{3.43}$$

where T_{in} is the input (motor) torque in lb in, ω_w is the input (worm) speed in *rpm* and 63 000 is a constant for unit conversions. Similarly, the output HP can be calculated from

$$HP_{out} = \frac{T_{out}\omega_{out}}{63\,000} \tag{3.44}$$

where T_{out} is the gearbox output torque in *lb in* and ω_{out} is the gearbox output speed in *rpm*.

Catalogs also specify a *Thermal HP*, which defines the maximum horsepower that can be transmitted continuously (30 min or longer) [7]. This is based on the temperature rise of the oil in the unit. If the thermal HP rating is less than the mechanical HP rating, the unit should be chosen using the thermal rating. Most positioning applications are intermittent. Therefore, heat build up in the gearbox is usually not an issue [9].

Some catalogs specify HP_{in} and HP_{out} but not the efficiency η_{GB} of the gearbox. Then, the efficiency can be found from

$$\eta_{GB} = \frac{HP_{out}}{HP_{in}} \tag{3.45}$$

Finally, the input and output torques of the gearbox are related through

$$T_{out} = \eta_{GB}T_{in}N_{GB} \tag{3.46}$$

3.10   Servo Motor and Gearhead Selection

In many systems, it is not possible to use direct drive due to load inertia, speed or size limitations. In such cases, it is necessary to select a motor and a gearhead to drive the load as shown in Figure 3.32. The gearhead increases the motor torque applied to the load by a factor of the gear ratio. But the load speed goes down by the same factor. There is a trade-off between speed and torque.

In this axis configuration, the gearhead input shaft speed is the same as the motor speed ω_m. Similarly, the gearhead output shaft speed is $\omega_{GB.out}$ is the same as the load speed ω_{load}. Gearhead manufacturers specify limits on the nominal and peak *output* torques that a gearhead can handle (T_{2N} and T_{2B}). In addition, there are limits on the gearhead *input* shaft nominal speed and maximum speed (n_{1N} and n_{1B}). Generally adapted notation among the manufacturers uses subscript "1" to indicate input and "2" to indicate output. Also, subscript "N" indicates nominal and "B" indicates maximum. For example, T_{2N} is the nominal output torque whereas n_{1B} is the maximum input speed. The load speed ω_{load} is also denoted as n_{2c} as shown in Figure 3.33.

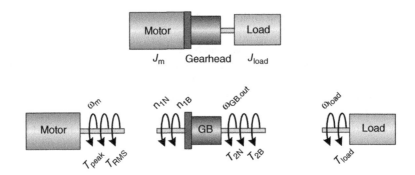

Figure 3.32   Motor coupled to the load via a gearhead

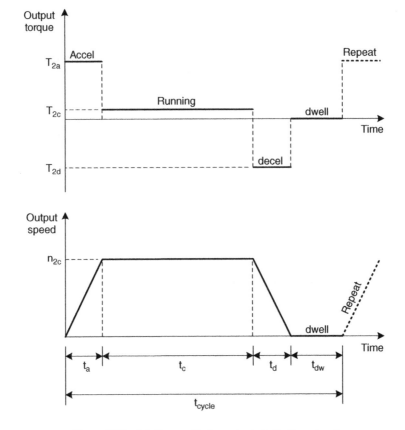

Figure 3.33   Motion profile for output shaft of gearhead

Section 3.4 explained how to calculate the peak and continuous (RMS) torques required for a particular motion profile for the load. Note that the formulas for the torques contain J_{total} and $T_{load \rightarrow M}$. Unless the axis is direct-drive as in Figure 3.3a, both J_{total} and $T_{load \rightarrow M}$ will involve the gearbox ratio N_{GB}. Therefore, proper gearhead selection is often an integral part of the motor selection and drive-train design.

When selecting a gearhead, we must ensure that its nominal and maximum output torque limits as well as its nominal and maximum input speeds are not exceeded when in service. To guarantee the rated service life of the gearhead, most manufacturers recommend operating it at or below its nominal output torque and nominal input speed limits. It is necessary to exceed these limits during brief acceleration or deceleration phases. This will not damage the gearhead as long as the maximum output torque limit is not exceeded.

There are different approaches to selecting a gearhead for an application. For example, after identifying a particular family of gearheads, first a gear ratio can be picked followed by torque and speed calculations to validate that the nominal and maximum limits are not exceeded with the given load and motion conditions of the application. If the limits are exceeded, another gear ratio is selected and the calculations are repeated. The method presented in this section has a slightly different approach. First, a range of possible gear ratios is identified. The upper limit of the range $N_{GB.upper}$ is found from the speed limitation imposed on the input shaft of the gearhead either by the servomotor rated speed or by the nominal input speed of the gearhead. The lower limit of the range is calculated from the inertia ratio J_R in Equations (3.19) and (3.20). Combining these equations and using the limiting case gives

$$\frac{J_{on\ motor\ shaft} + J_{load \to M} + J_{GB \to M}}{J_m} = 5$$

We can substitute

$$J_{load \to M} = \frac{J_{load}}{\eta_{GB} N_{GB}^2}$$

and solve for the gear ratio as

$$N_{GB.lower} = \sqrt{\frac{J_{load}}{\eta_{GB}(5J_m - J_{GB \to M} - J_{on\ motor\ shaft})}} \tag{3.47}$$

where $J_{on\ motor\ shaft}$ is the inertia on the motor shaft, such as a coupling between the motor and the gearhead. $J_{GB \to M}$ is the gearhead inertia reflected to its input shaft, which is how it is provided in catalogs. As explained in the selection procedure, if a desired motor inertia J_m is specified, Equation (3.47) can be used to calculate the lower limit for the necessary gear ratio to satisfy the inertia ratio criterion. Gearheads are not available in any gear ratio. Instead, they come in standard ratios such as 3, 4, 5, 7, 10, 12, etc. Once the range of possible ratios is identified, a standard gear ratio N_{GB} from a gearhead family can be selected in the range as

$$N_{GB.lower} \le N_{GB} \le N_{GB.upper}$$

A comprehensive gearhead selection analysis involves verification of whether the selected gearhead can support the axial and radial forces applied to its output shaft by the load. These steps were not considered here and should be incorporated into the design process, if these types of forces on the gearhead shaft are critical in a specific application. Backlash may be another important consideration in selecting a gearhead in certain motion control applications. Manufacturers offer gearheads with various levels of precision and backlash capabilities.

SERVO MOTOR AND GEARHEAD SELECTION PROCEDURE

GEARHEAD SELECTION

Gearhead selection is done by starting from the load and working towards the input shaft of the gearhead. Identify a potential product family of motors and a *compatible* family of gearheads. Note the rated speed ω_R for the motor with the smallest inertia, which is usually in the 3000–5000 rpm range for a typical servomotor.

1. Calculate the mean *output* torque T_{2m} for the gearhead. The angular acceleration of the output shaft can be found from the velocity profile in Figure 3.33 as $\ddot{\theta}_{\text{GB.out}} = n_{2c}/t_a$. Also, the output torques can be found using a similar approach as in Section 3.4.

$$T_{2a} = J_{\text{load}}\, \ddot{\theta}_{\text{GB.out}} + T_{\text{load}}$$

$$T_{2c} = T_{\text{load}}$$

$$T_{2d} = -J_{\text{load}}\, \ddot{\theta}_{\text{GB.out}} + T_{\text{load}}$$

$$T_{2m} = \sqrt[3]{\frac{n_{2a}t_a T_{2a}^3 + n_{2c}t_c T_{2c}^3 + n_{2d}t_d T_{2d}^3}{n_{2a}t_a + n_{2c}t_c + n_{2d}t_d}}$$

where $n_{2a} = n_{2d} = n_{2c}/2$ are the average output shaft speeds during acceleration and deceleration.

2. Calculate the duty cycle *ED%* of the motion to determine if the motion is continuous or cyclical [15].

$$ED\% = \frac{t_a + t_c + t_d}{t_{\text{cycle}}} \times 100$$

If $ED < 60\%$ and $(t_a + t_c + t_d) < 20\,\text{min}$, then cyclical. Go to step 3

If $ED > 60\%$ or $(t_a + t_c + t_d) > 20\,\text{min}$, then continuous. Go to step 10

Cyclical Operation

3. Calculate the upper limit for the potential gear ratios. Pick the smallest of the nominal gearhead *input* speed n_{1N} and the motor rated speed ω_R as the available speed ω_{avail} and calculate

$$N_{\text{GB.upper}} = \frac{\omega_{\text{avail}}}{n_{2c}}$$

SERVO MOTOR AND GEARHEAD SELECTION PROCEDURE (CONTINUED)

4. Calculate an *initial estimate* for the lower limit of the potential gear ratios using Equation (3.47). Ignore the gearhead inertia and efficiency. Use the motor with the smallest inertia J_m from the motor family.

$$N_{\text{GB.lower}}^{\text{est}} = \sqrt{\frac{J_{\text{load}}}{(5J_m - J_{\text{on motor shaft}})}}$$

5. Update the initial estimate by calculating the *actual lower limit* for potential gear ratios $N_{\text{GB.lower}}$. First, select a standard gearhead ratio closest to but bigger than $N_{\text{GB.lower}}^{\text{est}}$. Then, obtain the efficiency η_{GB} and inertia J_{GB} for this gearhead from the catalog. Use these values in Equation (3.47) to calculate the actual lower limit $N_{\text{GB.lower}}$.

6. Identify all possible standard gear ratios N_{GB} in the range

$$N_{\text{GB.lower}} \leq N_{\text{GB1}}, N_{\text{GB2}}, ..., N_{\text{GB}n} \leq N_{\text{GB.upper}}$$

Select one of them. A gear ratio at the upper end of the range will make the gearhead input speed closer to the nominal input speed. But it will also reduce the inertia ratio, which will make the system dynamically more responsive. Obtain T_{2N}, T_{2B}, n_{1N}, the gearhead inertia J_{GB} and efficiency η_{GB} for the selected gearhead from the catalog.

It may not always be possible to find standard gear ratios in the calculated range. If this is the case, then using the next size motor with bigger inertia may reduce the lower limit. If the gearhead was the limiting factor in step 3, the upper limit can be increased by using another gearhead family with higher nominal input speeds. Repeat steps 3 through 6 until a standard ratio can be selected.

7. Calculate service factor SF, where $C_{\text{h}} = 3600/t_{\text{cycle}}$ is the number of cycles per hour (t_{cycle} in seconds) [15].

SF	C_{h}
1.0	$C_{\text{h}} < 1000$
1.1	$1000 \leq C_{\text{h}} < 1500$
1.3	$1500 \leq C_{\text{h}} < 2000$
1.6	$2000 \leq C_{\text{h}} < 3000$
2.0	$3000 \leq C_{\text{h}}$

8. Calculate the maximum output torque for the application using larger of T_{2a} or T_{2d}.

$$T_{2\max} = \begin{cases} \eta_{\text{GB}} \cdot \text{SF} \cdot |T_{2a}| & \text{if } |T_{2a}| > |T_{2d}| \\ \eta_{\text{GB}} \cdot \text{SF} \cdot |T_{2d}| & \text{if } |T_{2a}| < |T_{2d}| \end{cases}$$

Continued

9. Check if the selected gearhead (with N_{GB}) can handle the mean and maximum output torques. Accept the selected gearhead if

$$T_{2\max} < T_{2B} \text{ AND } T_{2m} < T_{2N}$$

and go to step 16 for servo motor selection. If this condition is not met, then select a larger gearhead (bigger T_{2N}, T_{2B}) and repeat steps 8 and 9.

Continuous Operation
10. Calculate the mean output speed n_{2m} for the gearhead from

$$n_{2m} = \frac{n_{2a}t_a + n_{2c}t_c + n_{2d}t_d}{t_a + t_c + t_d}$$

11. Calculate the upper limit for the potential gear ratios. Pick the smallest of the nominal gearhead *input* speed n_{1N} and the motor rated speed ω_R as the available speed ω_{avail} and calculate

$$N_{GB.upper} = \frac{\omega_{avail}}{n_{2m}}$$

12. Calculate an *initial estimate* for the lower limit of the potential gear ratios using Equation (3.47). Ignore the gearhead inertia and efficiency. Use the motor with the smallest inertia J_m from the motor family.

$$N^{est}_{GB.lower} = \sqrt{\frac{J_{load}}{(5J_m - J_{on\,motor\,shaft})}}$$

13. Update the initial estimate by calculating the *actual lower limit* for potential gear ratios $N_{GB.lower}$. First, select a standard gearhead ratio closest to but bigger than $N^{est}_{GB.lower}$. Then, obtain the efficiency η_{GB} and inertia J_{GB} for this gearhead from the catalog. Use these values in Equation (3.47) to calculate the actual lower limit $N_{GB.lower}$.

14. Identify all possible standard gear ratios N_{GB} in the range

$$N_{GB.lower} \leq N_{GB1}, N_{GB2}, ..., N_{GBn} \leq N_{GB.upper}$$

Select one of them. A gear ratio at the upper end of the potential ratios will make the gearhead input speed closer to the nominal input speed. But it will also reduce the inertia ratio, which will make the system dynamically more responsive. Obtain T_{2N}, T_{2B}, n_{1N}, the gearhead inertia J_{GB} and efficiency η_{GB} for the selected gearhead from the catalog.

Continued

It may not always be possible to find standard gear ratios in the calculated range. If this is the case, then using the next size motor with bigger inertia may reduce the lower limit. If the gearhead was the limiting factor in step 11, the upper limit can be increased by using another gearhead family with higher nominal input speeds. Repeat steps 11 through 14 until a standard ratio can be selected.

15. Check if the selected gearhead (with N_{GB}) can handle the mean output torque. Accept the selected gearhead if

$$T_{2m} < T_{2N}$$

and go to step 16 for servo motor selection. If this condition is not met, then select a larger gearhead (bigger T_{2N}) and repeat this step.

MOTOR SELECTION

16. Calculate the T_{peak} and T_{RMS} that the motor should provide using the selected N_{GB} and J_m.
17. Make sure that the motor can deliver these torques. Accept the motor selection if $T_{peak} \leq T_{PR}$ and $T_{RMS} \leq T_{CR}$. If not, select the next motor with bigger torque output. If the motor inertia is different, then repeat steps 3 through 9 or steps 10 through 15 for gearhead selection and steps 16 and 17 for motor selection.
18. Compute the inertia ratio J_R using Equation (3.19) with the selected motor and gearhead combination. Accept the motor and gearhead combination if $J_R \leq 5$. Typically, $J_R \leq 5$ is used but this also depends on the desired system performance.

    If the inertia ratio criterion is not satisfied, then the motor inertia may be too small for the application. Pick the next size motor with bigger inertia. Also, increasing N_{GB} can significantly reduce the inertia ratio. If in steps 3 or 11 the limiting speed was the motor rated speed ω_R, then a faster motor can be tried. This may make the nominal gearhead input speed n_{1N} the limiting speed but it may allow selecting a bigger gear ratio. Steps 3 through 9 or steps 10 through 15 for gearhead selection and steps 16 through 18 for motor selection will need to be repeated.
19. Use the selected motor's torque–speed curves or data to check the speed and torque margins to make sure that the motor will be able to deliver the expected performance with some margin. In practice 30% margin for T_{RMS} and 50% margin for T_{peak} are common. It is desirable to have some margin (extra torque capability) for adjustments, if conditions change during the commissioning of the machine.

■ EXAMPLE 3.10.1

A small converting machine shown in Figure 3.34 is used to process sheet aluminum. The load inertia on the rewind axis is $J_{\text{load}} = 2 \times 10^{-2}$ kg-m^2 and the axis has 7 Nm load torque due to friction and tension. The motion uses trapezoidal velocity profile with $\omega_{\text{load}} = 150$ rpm. Change in the size of the rewind roll can be ignored. Select a NEMA 23 size servomotor and a gearhead to drive the rewind axis if the cycle consists of (a) $t_a = t_d = 30$ ms, $t_c = 3$ s, $t_{dw} = 4$ s, or (b) $t_a = t_d = 30$ ms, $t_c = 5$ s, $t_{dw} = 1$ s.

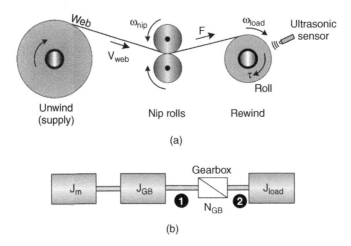

(a)

(b)

Figure 3.34 Motor and gearbox selection for a converting machine. (a) Sheet aluminum converting machine. (b) Rewind axis schematic

Solution

We will follow the procedure given in Section 3.10. NEMA 23 size AKM™ servomotors [20] by Kollmorgen and compatible EPL-X23 family of inline planetary gearheads [17] by GAM have been selected as a potential solution for the design. The AKM21C motor has the smallest inertia $J_m = 0.11 \times 10^{-4}$ kg-m^2. Therefore, it has been selected as an initial choice.

Part (a)

1. The angular acceleration of the gearhead output shaft (load) can be found from

$$\ddot{\theta}_{\text{GB.out}} = \frac{n_{2c}}{t_a}$$

$$= \frac{150}{0.030} \cdot \frac{2\pi}{60}$$

$$= 523.6 \, \text{rad/s}^2$$

The output torque during acceleration is

$$T_{2a} = J_{\text{load}} \, \ddot{\theta}_{\text{GB.out}} + T_{\text{load}}$$

$$= 2 \times 10^{-2} \cdot 523.6 + 7$$

$$= 17.47 \, \text{Nm}$$

Similarly, the output torque during running and deceleration phases are found as $T_{2c} = 7\,\mathrm{Nm}$ and $T_{2d} = -3.47\,\mathrm{Nm}$. Substituting these values along with $n_{2c} = 150, n_{2a} = n_{2d} = 75$ into the cubic root-mean formula gives $T_{2m} = 7.15\,\mathrm{Nm}$ as the mean torque on the output shaft of the gearhead.

2. The duty cycle of the motion is found from

$$ED\% = \frac{t_a + t_c + t_d}{t_{\text{cycle}}} \times 100$$

$$= \frac{0.03 + 3 + 0.03}{0.03 + 3 + 0.03 + 4} \times 100$$

$$= 43.3\%$$

which is considered cyclical motion. Hence, we continue with step 3 of the procedure in Section 3.10.

3. The gearhead catalog lists $n_{1N} = 3500\,\mathrm{rpm}$ as nominal input speed for the EPL-X23 family. The AKM21C motor has $\omega_R = 8000\,\mathrm{rpm}$ (operating at 320 VDC). Then, the upper limit for the potential gear ratios can be calculated from

$$N_{\text{GB.upper}} = \frac{\omega_{\text{avail}}}{n_{2c}}$$

$$= \frac{3500}{150}$$

$$= 23.33$$

4. An initial estimate of the lower limit for potential gear ratios can be found from

$$N_{\text{GB.lower}}^{\text{est}} = \sqrt{\frac{J_{\text{load}}}{(5J_m - J_{\text{on motor shaft}})}}$$

$$= \sqrt{\frac{2 \times 10^{-2}}{(5 \cdot 0.11 \times 10^{-4})}}$$

$$= 19.07$$

where $J_{\text{on motor shaft}} = 0$.

5. From the gearhead catalog, the closest standard ratio that is bigger than 19.07 is the 20:1 ratio. This unit has $\eta_{\text{GB}} = 0.92$ and $J_{\text{GB}} = 0.36 \times 10^{-4}\,\mathrm{kg\text{-}m^2}$. Using these values in Equation (3.47) we can calculate the lower limit for the potential gear ratios as

$$N_{\text{GB.lower}} = \sqrt{\frac{J_{\text{load}}}{\eta_{\text{GB}}(5J_m - J_{\text{GB}\to M} - J_{\text{on motor shaft}})}}$$

$$= \sqrt{\frac{2 \times 10^{-2}}{0.92(5 \cdot 0.11 \times 10^{-4} - 0.36 \times 10^{-4})}}$$

$$= 33.83$$

6. It is not possible to find a standard gear ratio since the lower and upper limits overlap and therefore do not define a range. This is an indication that the selected motor inertia is too small. Repeating steps 3 through 6 using inertia of the next bigger motor (AKM22C) leads to the same result.

   The same steps are repeated one more time with the next motor, which is AKM23D with $J_m = 0.22 \times 10^{-4}$ kg-m^2. The initial estimate for the lower limit is found as $N_{\text{GB.lower}}^{\text{est}} = 13.48$. Then, the range becomes

$$13.48 \leq N_{\text{GB1}}, N_{\text{GB2}}, ..., N_{\text{GB}n} \leq 23.33$$

In this range, gear ratios 16 or 20 are the only two standard options. The closest but bigger standard ratio 16:1 was selected. The efficiency $\eta_{\text{GB}} = 0.92$ and inertia $J_{\text{GB}} = 0.38 \times 10^{-4}$ kg-m^2 for this gearhead were obtained from the catalog.

   Finally, using these values in Equation (3.47) the actual lower limit was calculated as $N_{\text{GB.lower}} = 17.38$. The updated range was identified for possible standard gear ratios

$$17.38 \leq N_{\text{GB1}}, N_{\text{GB2}}, ..., N_{\text{GB}n} \leq 23.33$$

The only standard ratio available in this range is 20:1. Therefore, $N_{\text{GB}} = 20$ was selected and the following values were obtained from the catalog: $T_{2N} = 42$ Nm, $T_{2B} = 52$ Nm, $n_{1N} = 3500$ rpm, $J_{\text{GB}} = 0.36 \times 10^{-4}$ kg-m^2 and $\eta_{\text{GB}} = 0.92$.

   Even though a gear ratio has already been identified for the AKM23D motor, for the purpose of explaining the procedure, the next bigger motor (AKM24E) was also tried. The inertia for this motor is $J_m = 0.27 \times 10^{-4}$ kg-m^2. Repeating steps 3 through 6 with the AKM24E leads to the following range

$$14.97 \leq N_{\text{GB1}}, N_{\text{GB2}}, ..., N_{\text{GB}n} \leq 23.33$$

In this new range, there are both 16 and 20 standard ratios available. If 16, which is closer to the lower end of the range, is selected, the inertia ratio calculated in step 3.10 later becomes $J_R = 4.55$ and the motor spins at 2400 rpm. On the other hand, if 20, which is closer to the upper end of the range, is selected, the inertia ratio becomes $J_R = 3.35$ and the motor spins at 3000 rpm. The choice of 20:1 ratio with this motor makes the rewind axis more dynamically responsive due to the lower inertia ratio but the gearhead now runs at a speed closer to its nominal input speed, which may reduce its service life. We will continue the rest of the solution with the AKM23D motor and the 20:1 gear ratio.

7. The number of cycles per hour is $C_h = 3600/t_{\text{cycle}} = 3600/7.06 = 509.92$. Then, from the table we can obtain $SF = 1.0$.

8. Since $|T_{2a}| > |T_{2d}|$, the maximum output torque for the application is

$$T_{2\max} = \eta_{\text{GB}} \cdot SF \cdot |T_{2a}|$$

$$= 0.92 \cdot 1.0 \cdot 17.47$$

$$= 16.07 \, \text{Nm}$$

9. Since both $T_{2\max} < T_{2B}$ and $T_{2m} < T_{2N}$, the selected gearhead can handle the mean and maximum output torques required for the application. A spreadsheet shown in Figure 3.35 was prepared to implement the selection procedure.

Load Motion Profile			
Max. load speed	ωload (n2c) =	150	rpm
	=	15.71	rad/s
Acc. Time	ta =	0.03	sec
Run time	tc =	3	sec
Dec. time	td =	0.03	sec
Dwell time	tdw=	4	sec
Duty cycle	ED =	43.3	%
Operation mode		CYCLICAL	
Max. motor speed	ωm =	3000	rpm
		314.16	rad/s
Load accel	αload =	523.60	rad/s^2

Load			
Inertia	Jload =	2.00E-02	kg-m^2
Inertia on motor shaft	Jon_motor_shaft =	0.00E+00	kg-m^2
Load torque	Tload (Text) =	7.00	Nm

Motor			
Motor Inertia	Jm =	2.200E-05	kg-m^2
Motor rated speed	ωR =	5000	rpm

Gearhead			
Selected gearhead		GAM EPL-X23	
Selected gear ratio	NGB =	20	
GB efficiency	η =	0.92	
Nominal output torque	T2N =	42	Nm
Max. accel. output torque	T2B =	52	Nm
Nominal input speed	n1N =	3500	rpm
Max. input speed	n1B =	6000	rpm
Gearhead Inertia	JGB =	3.600E-05	kg-m^2

Mean Output Torque			
Accel. torque	T2a =	17.47	Nm
Running torque	T2c =	7.00	Nm
Decel. Torque	T2d =	-3.47	Nm
Mean output torque	T2m =	7.15	Nm

Cyclical Operation			
Available speed	ωavail =	3500	rpm
Est. lower limit for gear ratio	NGB_lower_est =	13.48	
Lower limit for gear ratio	NGB_lower =	17.14	
Upper limit for gear ratio	NGB_upper =	23.33	
Possible gear ratio range	17.14	<= NGB <=	23.33
Cycles per hour	Chr =	509.92	cyc/hr
Shock factor	SF =	1.0	
Max. output torque for app.	T2max =	16.07	Nm
	T2max < T2B ?	Y	
	T2m < T2N ?	Y	

Continuous Operation			
Mean output speed	n2m =		rpm
Available speed	ωavail =		rpm
Est. lower limit for gear ratio	NGB_lower_est =		
Lower limit for gear ratio	NGB_lower =		
Upper limit for gear ratio	NGB_upper =		
Possible gear ratio range			
	T2m < T2N ?		

Figure 3.35 First page of *Excel*® spreadsheet for cyclical operation calculations for part (a) in Example 3.10.1. Gear ratio 20:1 with AKM23D-BN servomotor

Next, we will select a servomotor. Since the motor selection starts at step 16 in the procedure, the following steps were numbered starting from 16 to match the steps in the procedure.

Servomotor Selection

16. We need to calculate T_{peak} and T_{RMS} torques that the motor should provide with the selected gear ratio. The peak torque can be found from Equation (3.37). The equation contains angular acceleration of the motor, which can be calculated from

$$\ddot{\theta}_m = N_{GB} \cdot \frac{\omega_{load}}{t_a}$$

$$= 10\,471.98 \, rad/s^2$$

where the load acceleration is found from the slope of the trapezoidal velocity profile. This was then converted into the motor acceleration by multiplying by the gear ratio.

The total inertia consists of the motor and gearhead inertia on the motor shaft plus the load inertia reflected to the motor shaft

$$J_{total} = J_m + J_{GB} + \frac{J_{load}}{\eta_{GB} N_{GB}^2}$$

$$= 1.12 \times 10^{-4} \, kg\text{-}m^2$$

where $J_{GB} = 3.6 \times 10^{-5} \, kg\text{-}m^2$ and $\eta_{GB} = 0.92$ were obtained from the manufacturer's catalog for the EPL-X23 gearhead with 20:1 ratio.

The 7 Nm load torque reflected to the motor shaft is

$$T_{load \to M} = \frac{T_{load}}{\eta_{GB} N_{GB}}$$

$$= 0.38 \, Nm$$

Then, from Equation (3.37), $T_{peak} = 1.557 \, Nm$. The running torque T_{run} can be found from Equation (3.38), which is simply equal to $T_{load \to M}$. The deceleration torque is found from Equation (3.39) as $T_{dec} = -0.796 \, Nm$. Finally, T_{RMS} is found from Equation (3.40) as $T_{RMS} = 0.273 \, Nm$. Figure 3.36 shows the second page of the spreadsheet for motor selection.

17. The AKM23D-BN motor has $T_{PR} = 42 \, Nm$ and $T_{CR} = 52 \, Nm$. Since $T_{peak} \leq T_{PR}$ and $T_{RMS} \leq T_{CR}$, the motor is accepted.

18. The inertia ratio is computed from Equation (3.19) as $J_R = 4.11$, which meets the criterion.

19. The peak torque margin can be calculated from

$$Peak \ Torque \ Margin = \frac{T_{peak}^m - T_{peak}^{app}}{T_{peak}^m}$$

where T_{peak}^m is the peak torque of the motor at the operating speed, T_{peak}^{app} is the peak torque required for the application. For the AKM23D-BN motor $T_{peak}^m = 3.84 \, Nm$. Using $T_{peak}^{app} =$

Motor Torques		
Motor accel	αm =	10471.98 rad/s^2
Total inertia	Jtotal =	1.12E-04 rad/s
Reflected load torque	Tload_m	0.380 Nm
Accel torque	Tacc =	1.557 Nm
Running torque	Trun =	0.380 Nm
Decel torque	Tdec =	-0.796 Nm
RMS torque	Trms =	0.273 Nm

Motor Selection		
Selected motor	Kollmorgen AKM23D-BN	
Peak torque	TPR =	3.84 Nm
Rated continuous torque	TCR =	1.03 Nm
Inertia ratio	JR =	4.11
	Tpeak < TPR ?	Y
	Trms < TCR ?	Y
	JR <= 5 ?	Y
Peak torque margin		59.5 %
Continuous torque margin		73.5 %
Speed margin		14.3 %

Figure 3.36 Second page of the *Excel*® spreadsheet for servomotor selection for part (a) in Example 3.10.1. Gear ratio 20:1, with AKM23D-BN servomotor

1.557 Nm, the peak torque margin is found as 59.5%. Similarly, the torque margin for the continuous torque is calculated as 73.5% from

$$\text{Continuous Torque Margin} = \frac{T^m_{RMS} - T^{app}_{RMS}}{T^m_{RMS}}$$

where $T^m_{RMS} = 1.03$ Nm and $T^{app}_{RMS} = 0.273$ Nm. Finally, the speed margin can be found from

$$\text{Speed Margin} = \frac{\omega_{avail} - \omega_{app}}{\omega_{avail}}$$

where ω_{app} is the operating motor speed of the application. In this case, the available speed is limited by the gearhead to $\omega_{avail} = 3500$ rpm and $\omega_{app} = 3000$ rpm. Then, the speed margin can be found as 14.3%.

Both the torque and speed margins leave extra capability for adjustments, if conditions change during the commissioning of the machine. This is similar to a factor of safety in the design. It allows adjustments to be made to the controller gains without saturating the motor and the drive, in case more speed or torque is needed later.

The selected products are the EPL-X23-020 gearhead by GAM and the AKM23D-BN motor (operating at 320 VDC) by Kollmorgen.

Part (b)
Duty cycle is calculated as

$$ED\% = \frac{t_a + t_c + t_d}{t_{cycle}} \times 100$$

$$= \frac{0.03 + 5 + 0.03}{0.03 + 5 + 0.03 + 1} \times 100$$

$$= 83.5\%$$

The machine is considered to be operating continuously. Hence, the following calculations start from step 10 of the procedure given in Section 3.10 to match the steps in the procedure.

10. The mean output speed for the gearhead can be found from

$$n_{2m} = \frac{n_{2a}t_a + n_{2c}t_c + n_{2d}t_d}{t_a + t_c + t_d}$$

$$= \frac{75 \cdot 0.03 + 150 \cdot 5 + 75 \cdot 0.03}{0.03 + 5 + 0.03}$$

$$= 149.11 \, \text{rpm}$$

11. From the catalog, the gearhead nominal input speed is $n_{1N} = 3500 \, \text{rpm}$. The AKM21C motor rated speed is $\omega_R = 8000 \, \text{rpm}$. Therefore, $\omega_{\text{avail}} = 3500 \, \text{rpm}$. Then,

$$N_{GB} = \frac{\omega_{\text{avail}}}{n_{2m}}$$

$$= \frac{3500}{149.11}$$

$$= 23.47$$

12. From the analysis in part (a), we know that the motor with the smallest inertia (AKM21C) and the next bigger one (AKM22C) did not produce a viable range for standard gear ratios. Using the AKM23D motor, an initial estimate of the lower limit for potential gear ratios can be found as $N_{GB.lower}^{est} = 13.48$.
13. The closest but bigger standard ratio is 16:1. The efficiency $\eta_{GB} = 0.92$ and inertia $J_{GB} = 0.38 \times 10^{-4} \, \text{kg-m}^2$ for this gearhead were obtained from the catalog. Finally, using these values in Equation (3.47) the lower limit was calculated as $N_{GB.lower} = 17.38$.
14. The following range was identified for possible standard gear ratios:

$$17.38 \leq N_{GB1}, N_{GB2}, ..., N_{GBn} \leq 23.47$$

The only standard ratio available in this range is 20:1. Therefore, $N_{GB} = 20$ was selected and the following values were obtained from the catalog: $T_{2N} = 42 \, \text{Nm}$, $T_{2B} = 52 \, \text{Nm}$, $n_{1N} = 3500 \, \text{rpm}$, $J_{GB} = 0.36 \times 10^{-4} \, \text{kg-m}^2$ and $\eta_{GB} = 0.92$.
15. The selected gearhead can handle the mean output torque since $T_{2m} < T_{2N}$. Hence, it is accepted.

Servomotor Selection

16. Calculating the peak and RMS motor torques with $N_{GB} = 20$ and $J_m = 0.22 \times 10^{-4} \, \text{kg-m}^2$ gives $T_{\text{peak}} = 1.557 \, \text{Nm}$ and $T_{\text{RMS}} = 0.367 \, \text{Nm}$.
17. The AKM23D-BN motor has $T_{PR} = 42 \, \text{Nm}$ and $T_{CR} = 52 \, \text{Nm}$. Since $T_{\text{peak}} \leq T_{PR}$ and $T_{\text{RMS}} \leq T_{CR}$, the motor is accepted.
18. The inertia ratio is computed from Equation (3.19) as $J_R = 4.11$, which meets the criterion.

19. The peak and continuous torque margins were computed as 59.5% and 64.4%, respectively. The speed margin is 14.3%. The continuous torque margin dropped to 64.4% compared to the 73.5% in the cyclical operation case. This is expected since in the continuous operation mode the machine has a longer run time, which increased the T_{RMS} required from the motor.

The selected products are the EPL-X23-020 gearhead by GAM and the AKM23D-BN motor (operating at 320 VDC) by Kollmorgen.

3.11   AC Induction Motor and Gearbox Selection

Selection of a vector-duty AC induction motor with a compatible gearbox follows a similar process to the selection of an AC servomotor and a servo gearhead in Section 3.10. The motor must be interfaced to a vector-control drive with proper current capacity to support the continuous and peak motor currents. Motor data from the catalog can be used directly or to construct a torque–speed curve as shown in Figure 3.26b.

A worm gear speed reducer (gearbox) compatible with the motor frame must be properly sized by applying the necessary service factors as recommended by the manufacturer. If the motor horsepower is known, then the gearbox can be selected based on input ratings. However, in drive-train design for motion control, often the driven load is known. Therefore, the gearbox is selected based on its output torque capacity [6].

In the axis configuration shown in Figure 3.37, the reducer input shaft speed is the same as the motor speed (ω_m). The input speed is often called the *worm RPM* in the catalogs. The reducer output shaft speed is $\omega_{GB.out}$. Manufacturers impose limits on the the peak and nominal output torques that the unit can handle ($O.T.^{acc}$ and $O.T.^{run}$). In addition, the input shaft nominal (continuous) speed is limited ($\omega_{GB.nom}$).

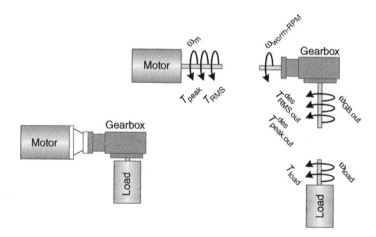

Figure 3.37   Induction motor coupled to the load via a gearbox

VECTOR-DUTY AC INDUCTION MOTOR AND GEARBOX SELECTION PROCEDURE

GEARBOX SELECTION

1. Calculate the gearbox *output* speed $\omega_{GB.out}$ required for the application. This will require analysis of the motion profile for the application and the kinematics of the transmission mechanism used in the drive-train design.
2. Identify a potential product family of vector-duty motors and a *compatible* family of gearboxes. Typically, worm gear speed reducers are used with these motors.
3. Divide the full-load speed of the motor at base frequency $\omega_{F.L.}$ by the gearbox *output* speed $\omega_{GB.out}$ required for the application to find the necessary gear ratio. Gearboxes come with preset ratios. Pick the closest possible gear ratio N_{GB} available in the family.
4. Calculate $T_{peak.out}$ and $T_{RMS.out}$ required at the *output shaft of the gearbox* to meet the demands of the motion profile and the load. Note that the calculation excludes the motor and gearbox since these are torques at the *output* of the gearbox to drive the load.
5. Determine the applicable service factor(s) based on the operating conditions of the axis using the data given in the gearbox catalog. Multiply $T_{peak.out}$ and $T_{RMS.out}$ by the service factor(s) to determine the "design" torques $T_{peak.out}^{des}$ and $T_{RMS.out}^{des}$. If the service factor is greater than 1.0, then the $T_{peak.out}$ and $T_{RMS.out}$ will be increased leading to a bigger gearbox selection to improve the wear life of the unit.
6. Determine the gearbox size using the HP and/or output torque ratings tables in the gearbox catalog. Using the motor full-load base speed (worm RPM) and the ratio N_{GB}, find the output torque (O.T.) limits of the smallest size gearbox in the tables. The output torque limits of the gearbox must be more than the design torques for the application calculated in step 5

$$T_{RMS.out}^{des} \leq O.T.^{run} \text{ AND } T_{peak.out}^{des} \leq O.T.^{acc}$$

where $O.T.^{acc}$ and $O.T.^{run}$ are the acceleration (peak) and nominal (continuous) torque output limits of the gearbox, respectively. If the torque limits of the gearbox are not exceeded, the selected gearbox is accepted. If they are exceeded, repeat step 6 with the next bigger size gearbox.

VECTOR-DUTY AC INDUCTION MOTOR AND GEARBOX SELECTION PROCEDURE
(CONTINUED)

---

7. If the load is connected to the gearbox using a sprocket or pulley, the overhung load ratings of the gearbox must be checked. Calculate the overhung load using Equation (3.42) and the $T^{\text{des}}_{\text{RMS.out}}$. Use the appropriate overhung load factor from the catalog. The applied load must be less than the overhung load rating of the selected gearbox.

MOTOR SELECTION

8. Calculate the T_{peak} and T_{RMS} that the *motor* should provide using the selected N_{GB}. At this point the motor is unknown. Therefore, use $J_m = 0$ in the torque calculations for now. Note that the gearbox inertia J_{GB} and efficiency η_{GB} can be found from the catalog for the gearbox selected in step 6.

9. From the motor catalog pick the smallest possible motor that can provide the calculated torques. This will require using the motor data table or making a torque–speed curve for the motor.

10. Compute the inertia ratio (Equation (3.19)) with the motor and gearbox combination.

    Accept a motor + gearbox combination if $J_R \leq 5$. Typically, $J_R \leq 5$ is used but this also depends on the desired system performance. Use an appropriate target inertia ratio based on the expected performance of the axis.

    If the inertia ratio criterion ($J_R \leq 5$) fails, then the motor inertia may be too small for the application. Pick the next size motor with bigger inertia and repeat step 10.

11. If the selected motor is accepted, then recalculate T_{peak} and T_{RMS} with the selected motor's inertia J_m. Check the gearbox output torque limits as in step 6 to verify that the gearbox is still within its torque output limits. If the gearbox limits are exceeded, then the next size gearbox should be used and all calculations in steps 8 through 11 should be repeated.

12. Use the selected motor's torque–speed curves or data to check the speed and torque margins to make sure that the motor will be able to deliver the expected performance with some margin. In practice 30% margin for T_{RMS} and 50% margin for T_{peak} are common. It is desirable to have some margin (extra torque capability) for adjustments, if conditions change during the commissioning of the machine.

■ EXAMPLE 3.11.1

The machine shown in Figure 3.38 retrieves a plastic cap from a magazine and inserts it into the end of the core of a roll. The turret has vacuum chucks and pneumatic pistons at both ends to hold and insert the caps. First, the turret retrieves a cap from the magazine. Then, it rotates 180° and stops to position the cap in front of the core. Next, the piston inserts the cap into the core and the vacuum chuck releases the cap. Finally, the turret makes another 180° rotation back to the magazine.

The turret inertia is 40 lb-ft². It follows a triangular velocity profile to make each 180° move in 0.8 s. The friction torque is 0.23 lb ft. Select a vector-duty AC induction motor and gearbox for the turret axis assuming no limitations imposed by the drive current capacity.

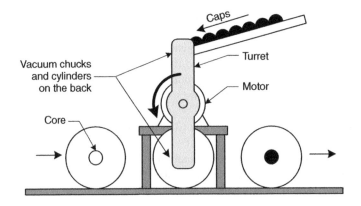

Figure 3.38 Core capping machine

Solution
We will follow the procedure given in Section 3.11.

Gearbox selection:

1. The output of the gearbox is directly connected to the turret, which will follow a triangular velocity profile ($t_m = 0$, $t_a = t_d = 0.4$ s). Therefore, from the area under the triangle, we can find the maximum gearbox output speed as

$$\omega_{\text{GB.out}} = \frac{S_{\text{max}}}{t_a}$$

$$= \frac{180}{0.4} \cdot \left(\frac{60}{360}\right)$$

$$= 75\,\text{rpm}$$

The 60/360 factor is for unit conversion into rpm.

2. In this design, we will use a Black Max® 1000:1 vector-duty motor [21] by Marathon™-Motors and a RG Series gearbox [9] by Cone Drive Operations, Inc.

3. The necessary gear ratio can be found from

$$N_{GB} = \frac{\omega_{F.L.}}{\omega_{GB.out}}$$

$$= \frac{1750}{75}$$

$$= 23.33$$

From the gearbox catalog, the standard ratio of 20:1 is selected. This will make the motor spin at 1500 rpm.

4. Torque calculations require the angular acceleration of the turret

$$\ddot{\theta}_{out} = \frac{\omega_{GB.out}}{t_a}$$

$$= \frac{75}{0.4}$$

$$= 187.5 \, \text{rpm/s}$$

The peak (acc) and RMS torques at the output of the gearbox can be found from

$$T_{acc.out} = J_{load} \cdot \ddot{\theta}_{out} + T_{fric}$$

$$= \frac{40 \cdot 187.5}{308} + 0.23$$

$$= 24.58 \, \text{lbft}$$

$$T_{run.out} = T_{fric}$$

$$= 0.23 \, \text{lb ft}$$

$$T_{dec.out} = -J_{load} \cdot \ddot{\theta}_{out} + T_{fric}$$

$$= -\frac{40 \cdot 187.5}{308} + 0.23$$

$$= -24.12 \, \text{lbft}$$

where the 308 factor is for unit conversion to obtain the torque in lb ft.

$$T_{RMS.out} = \sqrt{\frac{T_{acc.out}^2 \cdot t_a + T_{run.out}^2 \cdot t_m + T_{dec.out}^2 \cdot t_d}{t_a + t_m + t_d}}$$

$$= 24.35 \, \text{lb ft}$$

5. The manufacturer specifies two service factors for cycle shock and speed. It is assumed that the machine will have 1000 start/stop cycles per hour. Therefore, from the catalog $SF_{shock} = 1.0$. The motor speed with the 20:1 gear ratio will be 1500 rpm. Using the table in the catalog and linear interpolation, $SF_{speed} = 0.9$ was calculated. The overall service factor is $SF = (1.0) \cdot (0.9) = 0.9$.

The design torques can then be found as

$$T_{peak.out}^{des} = T_{acc.out} \cdot SF$$

$$= 22.12 \, \text{lb ft}$$

$$T_{\text{RMS.out}}^{\text{des}} = T_{\text{RMS.out}} \cdot SF$$

$$= 21.92 \, \text{lb ft}$$

6. The gearbox size can be determined form the catalog using the selected gear ratio and the gearbox output torques required by the application. Using the 20:1 ratio, $T_{\text{peak.out}}^{\text{des}}$ and $T_{\text{RMS.out}}^{\text{des}}$, we can select a model RG size 15 gearbox since it has $O.T.^{\text{run}} = 422/12 = 35.17$ lb ft and $O.T.^{\text{acc}} = 525/12 = 43.8$ lb ft, which are both more than the design torques.

7. The turret is directly connected to the output of the gearbox. Therefore, there is no overhung load to consider.

Motor Selection

8. The selected gearbox has an inertia of $J_{\text{GB}} = 11.4 \times 10^{-4}$ lb-in-s^2 reflected to its input shaft. The catalog specifies 72% efficiency for a worm speed of 100 rpm, which is closest to this application. Since at this point the motor is unknown, $J_m = 0$. The motor angular acceleration can be found as

$$\ddot{\theta}_m = \ddot{\theta}_{\text{out}} \cdot N_{\text{GB}}$$

$$= 3750 \, \text{rpm/s}$$

The total inertia is

$$J_{\text{total}} = J_m + J_{\text{GB}} + J_{\text{ref}}$$

$$= 0 + 11.4 \times 10^{-4} \cdot \left(\frac{32.2}{12}\right) + \frac{1}{0.72 \cdot 20^2} \cdot 40$$

$$= 0.142 \, \text{lb-ft}^2$$

where the $(32.2/12)$ factor is for unit conversion from lb-in-s^2 into lb-ft^2. The peak and RMS torques required from the motor can be found as

$$T_{\text{peak}} = J_{\text{total}} \cdot \ddot{\theta}_m + \frac{1}{\eta_{\text{GB}} N_{\text{GB}}} \cdot T_{\text{fric}}$$

$$= \frac{0.142 \cdot 3750}{308} + \frac{1}{0.72 \cdot 20} \cdot 0.23$$

$$= 1.73 \, \text{lb ft}$$

$$T_{\text{run}} = \frac{1}{\eta_{\text{GB}} N_{\text{GB}}} \cdot T_{\text{fric}}$$

$$= 7.99 \times 10^{-4} \, \text{lb ft}$$

$$T_{\text{dec}} = -J_{\text{total}} \cdot \ddot{\theta}_m + \frac{1}{\eta_{\text{GB}} N_{\text{GB}}} \cdot T_{\text{fric}}$$

$$= -1.73 \, \text{lb ft}$$

$$T_{\text{RMS}} = \sqrt{\frac{T_{\text{peak}}^2 \cdot t_a + T_{\text{run}}^2 \cdot t_m + T_{\text{dec}}^2 \cdot t_d}{t_a + t_m + t_d}}$$

$$= 1.73 \, \text{lb ft}$$

9. From the motor catalog, we can select a 1.5 HP Black Max® motor with 4.5 lb ft full-load torque and 0.140 lb-ft² inertia.

10. The inertia ratio can be found as

$$J_R = \frac{J_{GB} + J_{ref}}{J_m}$$

$$= 1.014$$

which is acceptable since $J_R \leq 5$.

11. Using the equations in step 8, the peak and RMS torques are recalculated with the inertia of the selected motor as $T_{peak} = 3.44$ lb ft and $T_{RMS} = 3.43$ lb ft. The resulting gearbox peak output torque can be calculated from

$$T_{peak.out}^{des} = \eta_{GB} \cdot N_{GB} \cdot T_{peak} \cdot SF$$

$$= 0.72 \cdot 20 \cdot 3.44 \cdot 0.95$$

$$= 47.06 \, lb \, ft$$

Similarly, $T_{RMS.out}^{des} = 46.92$ lb ft. Unfortunately, the gearbox output torque limits are exceeded since $O.T.^{run} = 35.17$ lb ft and $O.T.^{acc} = 43.8$ lb ft. Hence, we need to try a bigger size gearbox.

Next, Model RG size 25 gearbox is selected with $O.T.^{run} = 1411/12 = 117.6$ lb ft, $O.T.^{acc} = 1920/12 = 160$ lb ft, $\eta_{GB} = 0.75$ and $J_{GB} = 32.2 \times 10^{-4}$ lb-in-s². Steps 8 and 10 are repeated resulting in $J_R = 1.014$, $T_{peak} = 3.43$ lb ft and $T_{RMS} = 3.43$ lb ft. This selection is acceptable since the gearbox output torque limits are not exceeded and the motor can supply the required RMS torque. However, continuous torque margin is only 24%. It is desirable to have about 30% or more margin.

The next size motor (2 HP) is selected with 6.0 lb ft full-load torque and 0.130 lb-ft² inertia. This brings the total system inertia on the motor shaft to $J_{tot} = 0.272$ lb-ft². Steps 8 and 10 are repeated with the 2 HP motor and the size 25 gearbox resulting in $J_R = 1.092$, $T_{peak} = 3.31$ lb ft and $T_{RMS} = 3.31$ lb ft. Gearbox output limits are not exceeded.

12. Torque margins are computed as

$$Peak \, Torque \, Margin = \frac{T_{peak}^m - T_{peak}^{app}}{T_{peak}^m}$$

$$= \frac{9 - 3.31}{9}$$

$$= 63.2\%$$

$$RMS \, Torque \, Margin = \frac{T_{RMS}^m - T_{RMS}^{app}}{T_{RMS}^m}$$

$$= \frac{6 - 3.31}{6}$$

$$= 45\%$$

$T_{peak}^m = 9$ lb ft was used in the above torque margin calculations based on 150% of the full-load torque as the maximum available intermittent motor torque. The torque margins are acceptable. Hence, the selected 2 HP Black Max® vector duty induction motor and the Model RG size 25 gearbox with 20:1 gear ratio meet the design requirements.

3.12   Motor, Gearbox, and Transmission Mechanism Selection

This axis configuration shown in Figure 3.39 is obtained by connecting a motor and gearbox to one of the transmission mechanisms given in Section 3.3. The mechanism reflects the load torque, inertia and motion to its input shaft, which is now connected to the output shaft of the gearbox. The gearbox then reflects the motion, load inertia and torque to the motor connected to its input shaft.

The total inertia seen by the motor can be found from

$$J_{\text{total}} = J_m + J_{\text{GB}} + \frac{1}{\eta_{\text{GB}} N_{\text{GB}}^2} J_{\text{ref}}^{\text{trans}} \tag{3.48}$$

Similarly, the forces on the load are first reflected as load torque by the mechanism to its input shaft then by the gearbox to the motor shaft as

$$T_{\text{load}\to M} = \frac{1}{\eta_{\text{GB}} N_{\text{GB}}} T_{\text{load}\to in} \tag{3.49}$$

where J_{GB} is the gearbox inertia reflected to the motor shaft. $J_{\text{ref}}^{\text{trans}}$ is the inertia reflected by the mechanism to its input shaft (or the output shaft of the gearbox). The $T_{\text{load}\to in}$ is the load torque reflected by the mechanism to its input shaft (or the output shaft of the gearbox). Both of these terms are uniquely defined for each type of mechanism in Section 3.3.

Axis design can be started by first selecting a transmission type (lead screw, belt drive, rack-and-pinion, etc.). Then, select a mechanism from a catalog (such as [13, 28]). Next, substitute the appropriate $J_{\text{ref}}^{\text{trans}}$ equation from Section 3.3 into Equation (3.48) to calculate J_{total} for the axis. Similarly, use the appropriate equation for $T_{\text{load}\to in}$ for the selected

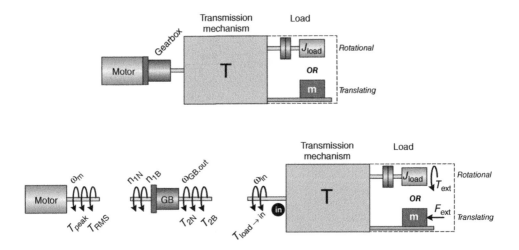

Figure 3.39   Motor coupled to the load via a gearbox and transmission mechanism

mechanism to calculate the $T_{load \to M}$. Now, the motor and gearbox selection can be completed following the procedure in Section 3.10 or Section 3.11 depending on the motor type.

There are now two ratios involved in the design (1) gearbox ratio, and (2) transmission ratio. By selecting a transmission mechanism first, and sizing a gearbox later, we are fixing one of these ratios to try to find the other. After following the procedure, if the result is not satisfactory, another type of mechanism or the same mechanism but with a different transmission ratio can be selected to repeat the gearbox/motor sizing procedure.

■ EXAMPLE 3.12.1

A machine needs to be designed to apply glue to a product which is 60 in long. The glue head must travel over the product at constant speed while applying the glue. Due to the structural design of the machine, there are only additional 4 in of space at each end of the 60 in linear travel to be used to accelerate and decelerate the glue head. The glue must be applied in 0.8 s. The glue head is 7 in wide and weighs 28 lb. Select components from the following product families: MKR-Series linear belt drive by Bosch Rexroth AG [32], AKM™ servomotor by Kollmorgen [20] and EPL-H gearhead by GAM [17].

Solution

The design requires selecting a linear belt drive transmission, gearbox and a motor. Following the procedure given in Section 3.12, we will first pick an appropriate linear belt drive from the linear actuator catalog. Then, we will select a gearbox and motor following the procedure in Section 3.10.

Linear Belt Drive Selection

The glue head is 7 in wide (178 mm). The MKR 15-65 or MKR 20-80 are two possible smallest linear modules that can carry the glue head since their carriage widths are both 190 mm. The MKR series has ball rail guideway and toothed belt. The maximum permissible drive torque for the MKR 15-65 and MKR 20-80 are 9.1 Nm and 32 Nm, respectively. Since the acceleration distance is very limited in this application, most likely a high acceleration torque will be required to move the glue head. Therefore, the MKR 20-80 was initially selected as the linear belt drive.

These modules can be ordered with custom length. The catalog provides a formula to calculate the necessary length for the module as

$$L = (S_{eff} + 2S_e) + 20 + L_{ca}$$

where S_{eff} is the effective stroke length, which is 60 in. S_e is the excess length at either end recommended for stopping distance, which is 4 in. L_{ca} is the width of the carriage, which is 190 mm for the MKR 20-80 unit. Note that the design specifications are given in U.S. customary units but the belt drive specifications are all in SI units. Therefore, all design specifications were first converted into SI units. The necessary module length is found as $L = 1937$ mm. The catalog also specifies a formula to calculate the inertia of the module reflected to its input as

$$J_{BeltDrive \to in} = (21.2 + 0.00379L) \times 10^{-4} \, \text{kg-m}^2$$

Substituting $L = 1937$ mm gives $J_{BeltDrive \to in} = 29 \times 10^{-4} \, \text{kg-m}^2$. This inertia is taken as the total reflected inertia of the belt drive. As explained in Example 3.8.1, the reflected inertia by the belt drive transmission must also include the load inertia as

$$
\begin{aligned}
J_{\text{ref}}^{\text{trans}} &= J_{BeltDrive \to in} + \frac{1}{\eta N_{BD}^2}\left(\frac{W_L + W_C}{g}\right) \\
&= 29 \times 10^{-4} + \frac{1}{0.95 \cdot 30.6^2}\left(\frac{28 \cdot 4.448 + 1.4 \cdot 9.81}{9.81}\right) \\
&= 17.1 \times 10^{-3}\,\text{kg-m}^2
\end{aligned}
$$

where the drive efficiency is $\eta = 0.95$ and the input pulley pitch diameter is $d_{ip} = 65.27 \times 10^{-3}$ m. Therefore, from Equation (3.31), $N_{BD} = 30.6\,(1/\text{m})$. Also, the 28 lb load was multiplying by 4.448 to convert it into Newtons.

From Equation (3.28) we can calculate the total external force. The weight of the carriage W_C was specified as 1.4 kg in the catalog. Assuming friction coefficient is $\mu = 0.5$, the external force is found as $F_{ext} = 69.14$ N. From Equation (3.33), we can calculate the load torque reflected to the input shaft of the belt drive module as $T_{\text{load} \to in} = 2.38$ Nm.

Gearhead Selection

At this point, we are ready to begin the selection of the gearbox and the AC servo motor following the procedure in Section 3.10. Note that $T_{\text{load} \to in}$ is the same as T_{load} and $J_{\text{ref}}^{\text{trans}}$ is the same as J_{load} in Section 3.10 since the input shaft of the belt drive is directly connected to the output shaft of the gearhead.

The GAM EPL-H size 84 gearhead [17] is available with a mounting interface that is compatible with the belt drive. Therefore, it was selected as an initial choice. The AKM4x family of motors [20] is compatible with this unit. The AKM41H motor has the smallest inertia $J_m = 0.81 \times 10^{-4}$ kg-m^2. Therefore, it has been selected as an initial choice.

1. We need to calculate the mean output torque for the gearhead. The calculation requires angular acceleration $\ddot{\theta}_{\text{GB.out}}$ of the output shaft.

   The gearhead output speed can be found from analyzing the motion profile. The axis must travel at constant speed for 60 inches in 0.8 s to uniformly apply the glue. Therefore, $V_{\text{max}} = 60/0.8 = 75$ in/s (1.9 m/s). This is the speed of the carriage. If we divide it by the radius of the input pulley of the belt drive, we can find the rotational speed needed at the input shaft of the belt drive, which is also the gearhead output running speed

$$
\begin{aligned}
n_{2c} &= \frac{1.9}{(65.27 \times 10^{-3})/2} \\
&= 58.37\,\text{rad/s} \\
&= 557.4\,\text{rpm}
\end{aligned}
$$

Then, the angular acceleration of the gearhead output shaft (load) can be found from

$$
\begin{aligned}
\ddot{\theta}_{\text{GB.out}} &= \frac{n_{2c}}{t_a} \\
&= 547.25\,\text{rad/s}^2
\end{aligned}
$$

The output torque during acceleration is

$$T_{2a} = J_{\text{load}}\,\ddot{\theta}_{\text{GB.out}} + T_{\text{load}}$$

$$= 17.1 \times 10^{-3} \cdot 547.25 + 2.38$$

$$= 11.73\,\text{Nm}$$

Similarly, the output torque during running and deceleration phases are found as $T_{2c} = 2.38\,\text{Nm}$ and $T_{2d} = -6.976\,\text{Nm}$. Substituting these values along with $n_{2c} = 557.4$, $n_{2a} = n_{2d} = 278.7$ into the cubic root-mean formula gives $T_{2m} = 4.426\,\text{Nm}$ as the mean torque on the output shaft of the gearhead.

2. By observation, duty cycle is 100% since there is no dwell time in the cycle of the machine. Using the formula, it can also be calculated as

$$ED\% = \frac{t_a + t_c + t_d}{t_{\text{cycle}}} \times 100$$

$$= \frac{0.107 + 0.8 + 0.107}{0.107 + 0.8 + 0.107} \times 100$$

$$= 100\%$$

The machine is operating continuously. Hence, the following calculations start from step 10 of the procedure given in Section 3.10 to match the steps in the procedure.

10. The mean output speed for the gearhead can be found from

$$n_{2m} = \frac{n_{2a}t_a + n_{2c}t_c + n_{2d}t_d}{t_a + t_c + t_d}$$

$$= 498.74\,\text{rpm}$$

11. From the catalog, the gearhead nominal input speed is $n_{1N} = 3000\,\text{rpm}$. The AKM41H motor rated speed is $\omega_R = 6000\,\text{rpm}$. Then, $\omega_{\text{avail}} = 3000\,\text{rpm}$ and

$$N_{\text{GB.upper}} = \frac{\omega_{\text{avail}}}{n_{2m}}$$

$$= \frac{3000}{498.74}$$

$$= 6.015$$

12. An initial estimate of the lower limit for potential gear ratios can be found from

$$N_{\text{GB.lower}}^{\text{est}} = \sqrt{\frac{J_{\text{load}}}{(5J_m - J_{\text{on motor shaft}})}}$$

$$= \sqrt{\frac{17.1 \times 10^{-3}}{(5 \cdot 0.81 \times 10^{-4})}}$$

$$= 6.496$$

where $J_{\text{on motor shaft}} = 0$.

13. It is not possible to find a standard gear ratio since the estimated lower and upper limits overlap and therefore do not define a range. This is an indication that the selected motor inertia is too small. Repeating steps 11 through 13 using inertia of the next bigger motor (AKM42E) leads to the same result.

   The same steps are repeated one more time with the next motor, which is AKM43L with $J_m = 2.1 \times 10^{-4}$ kg-m^2. The initial estimate for the lower limit is found as $N_{\text{GB.lower}}^{\text{est}} = 4.034$. The closest but bigger standard ratio is 5:1. Using $N_{\text{GB}} = 5$, efficiency $\eta_{\text{GB}} = 0.92$ and inertia $J_{\text{GB}} = 1.05 \times 10^{-4}$ kg-m^2 for this gearhead were obtained from the catalog. Finally, using these values in Equation (3.47) the actual lower limit was calculated as $N_{\text{GB.lower}} = 4.433$.

14. The following range is found

$$4.433 \leq N_{\text{GB}} \leq 6.015$$

Looking at the gearhead catalog, only the 5:1 standard ratio is in this range. Hence, $N_{\text{GB}} = 5$ is kept as the selection. This gearhead has $T_{2N} = 50$ Nm, $T_{2B} = 75$ Nm, $n_{1N} = 3000$ rpm, $J_{\text{GB}} = 1.05 \times 10^{-4}$ kg-m^2 and $\eta_{\text{GB}} = 0.92$.

   The gearhead input speed will be $n_{1c} = N_{\text{GB}} \cdot n_{2c} = 2787$ rpm, which is near but below the nominal input speed of the gearhead.

15. The selected gearhead can handle the mean output torque since $T_{2m} < T_{2N}$. Hence, it is accepted.

Servomotor Selection

16. Calculating the peak and RMS motor torques with $N_{\text{GB}} = 5$ and $J_m = 2.1 \times 10^{-4}$ kg-m^2 gives $T_{\text{peak}} = 3.41$ Nm and $T_{\text{RMS}} = 1.43$ Nm.

17. The AKM43L motor has $T_{\text{PR}} = 16.0$ Nm and $T_{\text{CR}} = 2.53$ Nm. Since $T_{\text{peak}} \leq T_{\text{PR}}$ and $T_{\text{RMS}} \leq T_{\text{CR}}$, the motor is accepted.

18. The inertia ratio is computed from Equation (3.19) as $J_R = 4.04$, which meets the criterion.

19. The peak and continuous torque margins were computed as 78.7% and 43.7%, respectively. The speed margin is 7.1%.

   The selected products are the EPL-H-084-005 gearhead by GAM, the AKM43L motor (operating at 320 VDC) by Kollmorgen, and the MKR 20-80 linear belt drive by Bosch Rexroth AG.

Problems

1. What is the gear ratio N_{GB} for each of the gearboxes shown in Figure 3.40?

2. The axis shown in Figure 3.40b has the following parameters: $n_1 = 30, n_2 = 60, n_3 = 30, n_4 = 90, J_{\text{load}} = 5 \times 10^{-4}$ kg-m^2 and $J_m = 3 \times 10^{-6}$ kg-m^2. The gearbox efficiency is 94%. What is the load inertia seen by the motor? What is the inertia ratio?

3. A pulley-and-belt transmission has a big and a small pulley. Figure 3.41 shows the cross-section of the pulleys and Table 3.3 provides the dimensions. If the pulleys are made out of steel with $\rho = 0.280$ lb/in^3 density, calculate the inertia of each pulley in lb-in-s^2.

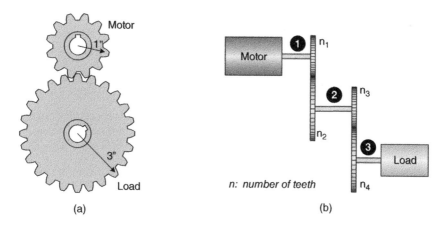

(a)   (b)

Figure 3.40   Problem 1. (a) Simple gearbox. (b) Gear train

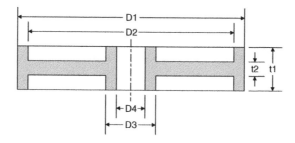

Figure 3.41   Pulley geometry for Problem 3

Table 3.3   Pulley dimensions (in) for Problem 3

Pulley	D1	D2	D3	D4	t1	t2
Big	5	4.5	1.125	0.625	1.5	0.25
small	2.5	2	1.125	0.625	1.5	0.25

4. The conveyor shown in Figure 3.42 has a drive roller with diameter D_{DR} and inertia J_{DR}, back roller with diameter D_{BR} and inertia J_{BR} and three idlers each with diameter D_{ID} and inertia J_{ID}. The weight of the conveyor belt is W_{Cbelt}. The drive roller has a big pulley with diameter D_{BP} and inertia J_{BP}. It is connected to a small pulley with diameter D_{SP} and inertia J_{SP} on the gearbox. The gearbox has N_{GB} gear ratio and J_{GB} inertia, which is already reflected to its input. The motor inertia is J_m.

   When the conveyor is carrying its maximum rated load W_L, what are the total inertia J_{total} on the motor shaft and the inertia ratio J_R? Ignore the weight of the drive pulley belt.

5. The drive and back rollers of the conveyor in Figure 3.42 have 6 in diameters. The three idlers have 2 in diameter each. All rollers are hollow cylinders with 0.25 in wall

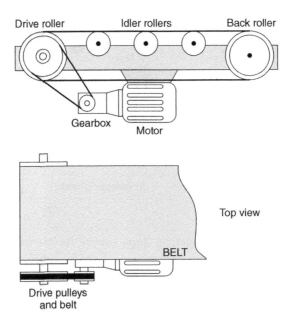

Figure 3.42 Conveyor for Problem 4

thickness, 18 in length and are made out of aluminum with $\rho = 0.096\,\text{lb/in}^3$ density. The conveyor belt weighs 18 lb. The drive pulleys are as specified in Problem 3 and as shown in Figure 3.42. The drive belt weight is negligible. The gearbox has 5:1 gear ratio and has $J_{GB} = 0.319\,\text{lb-in}^2$ inertia reflected to its input. The motor inertia is $J_m = 0.110\,\text{lb-ft}^2$. Both the conveyor and the gearbox have an efficiency of $\eta = 0.9$.

When the conveyor is carrying its maximum rated load of 500 lb, calculate the total inertia J_{total} on the motor shaft and the inertia ratio J_R. Ignore the weight of the drive pulley belt.

6. What are the peak and continuous torque margins and the speed margin for an X-axis motor of the gantry machine in Example 3.4.1?

7. Sealed boxes filled with a product are pushed into a bin from a conveyor as shown in Figure 3.43. The pusher was built using a ball screw linear transmission and a AC servomotor. The steel ball screw has $\rho = 0.280\,\text{lb/in}^3$ density, 16 mm pitch diameter, 533 mm length and 10 mm/rev lead. The carriage weighs 1.09 lb. The drive screw efficiency is 90%. The transmission coefficient of friction is $\mu_{\text{screw}} = 0.01$. The coefficient of friction between the box and the belt is $\mu_{\text{box}} = 0.4$. The motor inertia is $1.5 \times 10^{-4}\,\text{lb-in-s}^2$. Each box weighs 15 lb.

The pusher waits at 1 in away from the left edge of the conveyor. When a box arrives, the conveyor stops. The pusher accelerates over the distance of 1 in so that when it is at the left edge of the conveyor it reaches the speed of 18 in/s. Then, it travels at this speed for 1 second to push the box into the bin. When the pusher is at the right edge of the conveyor, it decelerates to stop at the same rate as its acceleration.

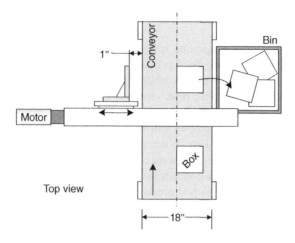

Figure 3.43   Box pusher for Problem 7

After the motion starts from the wait position, when does the pusher touch the box? Make a velocity versus time plot and a torque versus time plot for the entire motion of the axis.

8. A winding machine is used to make hardwound paper towel rolls (Figure 3.44) that are used in commercial restroom hand towel dispensers. The machine unwinds craft paper from a large parent roll, embosses it, slits it into individual sections and then rewinds it onto a cardboard core. The rewind axis produces paper logs with 8 in diameter and 6 ft length and 2 in hollow cardboard core. Each log is then cut into 9 rolls of paper towels that are 8 in wide.

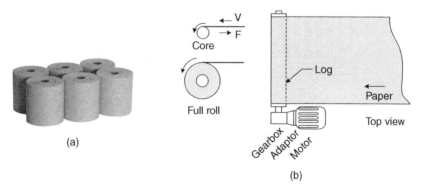

(a)

(b)

Figure 3.44   Rewind axis of a winding machine for Problem 8. (a) Hardwound paper towels. (b) Axis and motor

The machine operates with a web speed of 200 ft/min and the web tension is 1 lb/in of linear length of the paper. Select a three-phase AC Induction vector motor and a gearbox for the rewind axis from manufacturer catalogs (such as [7] and [21]). Hint: Consider the changing motor torque and speed requirements as the roll is wound while keeping the tension constant.

9. The outdated controller of a machine is being retrofitted with another controller that is already on hand. The new controller has many desirable features but its rated current outputs are slightly less than the old controller's current outputs. One of the axes on the machine has a motor whose rated peak torque is $T_{PR} = 4.38$ Nm, and rated continuous torque is $T_{CR} = 1.33$ Nm. The motor rated peak current is $I_{PR} = 5.3$ A, and rated continuous current is $I_{CR} = 1.6$ A. The new controller operates at the same voltage but provides only $I_{PR} = 4.0$ A, and $I_{CR} = 1.1$ A. When the motor is connected to the new controller, its torque values will be de-rated proportionally to the currents. What are the new values for the T_{PR} and T_{CR} for the new controller+motor combination?

10. A conveyor has 3 boxes of finished products on it. Each box weighs 20 kg. The two main rollers have $J_{roller} = 2.33 \times 10^{-3}$ kg-m^2 inertia each and $d_r = 8$ cm diameter. The conveyor has several idler rollers to support the weight on the belt but the idler inertia are negligible. The belt weighs 5 kg. The velocity profile of the desired motion is $t_a = t_d = 100$ ms, $t_m = 1$ s, the belt speed is $V_{max} = 0.5$ m/s. The belt friction coefficient is $\mu = 0.2$ and the conveyor efficiency is $\eta = 0.8$.

   (a) Calculate the T_{peak} and T_{RMS}.
   (b) Select a servomotor from the BSM63N family with IEC mountings [1] by ABB and a compatible planetary gearbox from the LP$^+$ Generation 3 family [36] by WITTEN-STEIN holding Corp.

11. A pallet dispenser shown in Figure 3.45 is used to dispense one pallet at a time to a conveyor from a stack of pallets. The machine has two axes, one for lifting the stack of pallets and the other for moving the forks in or out of the pallet stack. Initially, a stack of pallets is placed into the machine.

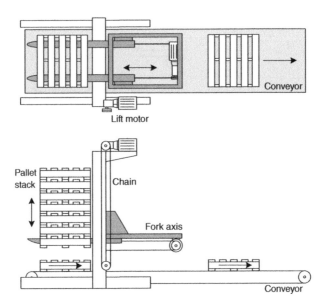

Figure 3.45   Pallet dispenser for Problem 11

The machine operates as follows: (1) The forks are moved into the pallet above the bottom pallet in the stack, (2) The machine lifts the stack leaving the bottom pallet on the conveyor, (3) The conveyor is started to dispense that pallet out of the machine, (4) The conveyor is stopped, the stack is lowered and placed on the conveyor, and (5) The forks are pulled out and raised to the level of the second pallet from the bottom. The cycle is repeated from step 1.

The lift axis needs to move 24 inches in 5 s ($t_a = t_d = 1$ s). It then dwells for 3 s. The machine can handle a stack of 10 pallets. The total weight of the pallets and the carriage is 1500 lb. The chain weighs 11 lb and uses two 3 in sprockets with $J_{sp} = 9.69 \times 10^{-4}$ lb-in-s^2 inertia each. The chain drive efficiency is 95%. Select a three-phase AC induction vector motor and a gearbox for the lift axis from manufacturer catalogs (such as [7] and [21]).

References

[1] ABB Corp. (2014) AC servo motors BSM Series. http://www.baldormotion.com/products/servomotors/n_series /bsm_nseries.asp (accessed 16 November 2014).

[2] Allied Motion Technologies, Inc. (2014) Quantum NEMA 23 Brushless Servo Motors. http://www .alliedmotion.com/Products/Series.aspx?s=51 (accessed 14 November 2014).

[3] Apex Dynamics USA. (2014) AB-Series High Precision Planetary Gearboxes. http://www.apexdynamicsusa .com/ab-series-high-precision-planetary-gearboxes.html (accessed 10 November 2014).

[4] Apex Dynamics USA. (2014) ABR-Series High Precision Planetary Gearboxes. http://www.apexdynamicsusa .com/products/abr-series-high-precision-planetary-gearboxes.html (accessed 10 November 2014).

[5] Apex Dynamics USA. (2014) PN-Series High Precision Planetary Gearboxes. http://www.apexdynamicsusa .com/pn-series-high-precision-planetary-gearboxes.html (accessed 10 November 2014).

[6] Baldor-Dodge. (2014) Tigear-2 Speed Reducers. http://www.dodge-pt.com/products/gearing/tigear2/index.html (accessed 2 September 2014).

[7] Cone Drive Operations, Inc. (2013) Cone Drive® Model HP Gearing Solutions (HP.0906.20). http://www .conedrive.com/library/userfiles/ModelHP-catalog-Printed.pdf (accessed 13 December 2013).

[8] Cone Drive Operations, Inc. (2014) Cone Drive Motion Control Solutions. http://conedrive.com/Products /Motion-Control-Solutions/motion-control-solutions.php (accessed 23 Februrary 2014).

[9] Cone Drive Operations, Inc. (2014) Cone Drive® Model RG Servo Drive. http://conedrive .com/Products/Motion-Control-Solutions/model-rg-gearheads.php (accessed 23 Februrary 2014).

[10] DieQua Corp. (2014) Manufacturer and Supplier of Gearboxes for Motion Control and Power Transmission. http://www.diequa.com/index.html (accessed 20 November 2014).

[11] George Ellis. *Control Systems Design Guide*. Elsevier Academic Press, Third edition, 2004.

[12] Emerson Industrial Automation. (2014) Servo Motors Product Data, Unimotor HD. http://www .emersonindustrial.com/en-EN/documentcenter/ControlTechniques/Brochures/CTA/BRO_SRVMTR_1107.pdf (accessed 14 November 2014).

[13] Festo Crop. (2014) EGC Electric Linear Axis with Toothed-belt. http://www.festo.com/cms/en-us_us /11939.htm (accessed 31 October 2014).

[14] Festo Crop. (2014) EGSK Electric Slide. http://www.festo.com/cat/de_de/data/doc_engb/PDF/EN/EGSK -EGSP_EN.PDF (accessed 31 October 2014).

[15] GAM. (2013) 2013 GAM Catalog. http://www.gamweb.com/documents/2013GAMCatalog.pdf (accessed 19 March 2013).

[16] GAM. (2014) Gear Reducers. http://www.gamweb.com/gear-reducers-main.html (accessed 21 January 2014).

[17] GAM. (2014) High Performance: EPL Series. http://www.gamweb.com/gear-reducers-high-precision-EPL.html (accessed 21 January 2014).

[18] Georgia-Pacific LLC. (2014) Envision® High Capacity Roll Paper Towel. http://catalog.gppro.com /catalog/6291/8373 (accessed 2 November 2014).

[19] Austin Hughes and Bill Drury. *Electric Motors and Drives; Fundamentals, Types and Applications*. Elsevier Ltd., Fourth edition, 2013.

[20] Kollmorgen. (2014)   Kollmorgen AKM Servomotor Selection Guide with AKD Servo Drive Systems. http://www.kollmorgen.com/en-us/products/motors/servo/akm-series/akm-series-ac-synchronous-motors/ac-synchronous-servo-motors/ (accessed 21 March 2014).

[21] Marathon Motors. (2013) SB371-Variable Speed Motor Catalog. http://www.marathonelectric.com/motors/index.jsp (accessed 25 October 2013).

[22] Marathon Motors. (2014) Three Phase Inverter (Vector) Duty Black Max® 1000:1 Constant Torque, TENV Motor. http://www.marathonelectric.com/motors/index.jsp (accessed 4 May 2014).

[23] John Mazurkiewicz. (2007) Motion Control Basics. http://www.motioncontrolonline.org/files/public/Motion_Control_Basics.pdf (accessed 7 October 2012.

[24] Robert Norton. *Machine Design: An Integrated Approach*. Prentice-Hall, 1998.

[25] Warren Osak. *Motion Control Made Simple*. Electromate Industrial Sales Limited, 1996.

[26] Parker Hannifin Corp. (2008)   ERV Series Rodless Actuators.   http://www.parkermotion.com/actuator/18942ERV.pdf (accessed 2 December 2014).

[27] Parker Hannifin Corp. (2013) BE Series Servo Motors. http://divapps.parker.com/divapps/emn/download/Motors/BEServoMotors_2013.pdf (accessed 23 July 2013).

[28] Parker Hannifin Corp. (2013) Parker 404XE Ballscrew Linear Actuators. http://www.axiscontrols.co.uk/ballscrew-linear-actuators (accessed 23 July 2013).

[29] Parker Hannifin Corp. (2013)   PE Series Gearheads: Power and Versatility in an Economical Package. http://www.parkermotion.com/gearheads/PV_Gearhead_Brochure.pdf (accessed 23 July 2013).

[30] Parker Hannifin Corp. (2014) Screw Driven Tables - XE Series. http://www.parkermotion.com/literature/precision_cd/CD-EM/daedal/cat/english/SectionB.pdf (accessed 6 June 2014).

[31] Parker Hannifin Corp. (2014) XE Economy Linear Positioners Catalog 2006/US. http://www.parkermotion.com/literature/precision_cd/CD-EM/daedal/cat/english/402XE-403XE%20print%20catalog%2012_8_06.pdf (accessed 6 June 2012).

[32] Rexroth Bosch AG. (2014) Bosch-Rexroth Linear Modules with Ball Rail Systems and Toothed Belt Drive (MKR). http://www.boschrexroth.com/en/xc/products/product-groups/linear-motion-technology/linear-motion-systems/linear-modules/index (accessed 20 May 2014).

[33] Rockford Ball Screw Co. (2014) ACME Screws. http://www.rockfordballscrew.com/products/acme-screws/ (accessed 5 December 2014).

[34] Rockford Ball Screw Co. (2014)   Ball Screws.   http://www.rockfordballscrew.com/products/ball-screws/ (accessed 5 December 2014).

[35] The Gates Corp. (2014) PowerGrip® Belt Drive. http://www.gates.com/products/industrial/industrial-belts (accessed 29 October 2014).

[36] WITTENSTEIN holding Corp. (2014)   LP+/LPB+ Generation 3 Planetary Gearheads.   http://www.wittenstein-us.com/planetary-gearhead-lp-plus-lpb-plus.php (accessed 26 November 2014).

[37] WITTENSTEIN holding Corp. (2014)   Precision Gearboxes.   http://www.wittenstein-us.com/Precision-Gearboxes.html (accessed 26 November 2014).

4

Electric Motors

Electric motors convert electrical energy into mechanical energy. According to the U.S. Department of Energy, more than 50% of all electrical energy in the United States is consumed by electric motors. Pumps, machine tools, fans, household appliances, disk drives, and power tools are just a few of the diverse applications of electric motors.

The most common motors used in industrial motion control are three-phase AC servo and induction motors. An AC servo motor has permanent magnets on its rotor. The stator contains three-phase windings. These motors are designed to keep the rotor inertia low. As a result, they have fast dynamic response making them very popular in motion control. They also have an integrated position sensor, such as an encoder, to allow precise control with a drive. An induction motor uses a simple and rugged rotor which is externally or internally shorted. The stator contains three-phase windings. The relatively low cost, rugged construction, and good performance made induction motors the most widely used motors in industry. In motion control, inverter-rated (vector-duty) induction motors are employed. These motors have specialized windings that can withstand the demands of the frequent starting and stopping. They are equipped with a position sensor and controlled by a vector-drive in applications such as rewind machines.

This chapter begins with the fundamental concepts such as electrical cycle, mechanical cycle, poles, and three-phase windings. Then, the generation of a rotating magnetic field by the stator is described. Hall sensors and six-step commutation are discussed. Next, construction details of AC servo motors are presented. Torque generation performance with sinusoidal and six-step commutation are compared. This is followed by an overview of the construction of a squirrel-cage AC induction motor. Constant speed operation with across-the-line power and variable speed operation with a variable frequency drive are discussed. The chapter concludes with mathematical and simulation models for both types of motors.

4.1 Underlying Concepts

In both types of motors that we will study in this chapter, the stator is made up of three-phase windings, which are shifted 120° from each other mechanically. Each phase winding is essentially an electromagnet. When the stator windings are energized, they create a rotating

Industrial Motion Control: Motor Selection, Drives, Controller Tuning, Applications, First Edition. Hakan Gürocak.
© 2016 John Wiley & Sons, Ltd. Published 2016 by John Wiley & Sons, Ltd.

magnetic field inside the motor. Note that the stator remains stationary, but the magnetic field generated by the windings rotates. The rotor, which is the moving part of the motor, is pulled by this rotating magnetic field and starts spinning.

An electromagnet can be constructed by winding a wire into a coil and applying current as shown in Figure 4.1. If the coil is wound on an iron core, the magnetic field becomes stronger. The magnetic flux lines create a North and a South Pole. The direction of the current determines the polarity of the electromagnet. If you wrap the fingers of your right hand along the current direction (conventional current) on the coil, your thumb will point in the direction of the North Pole of the electromagnet. If the current polarity is reversed, the magnetic poles are also reversed. The strength of the magnetic field can be increased by increasing the number of *coil turns* and/or the current.

A pair of windings can be used to create one *phase* of the stator as shown in Figure 4.2a. Each winding with its iron core is called a *pole*. The winding directions of the coils are such that when the current is applied to the phase, North and South Poles are created.

Assume that a bar magnet is pinned in the center of the stator. If the current is applied in the instant as shown in Figure 4.2b, the magnet will rotate clockwise to align with the phase windings since opposite poles on the magnet and the stator will attract each other. While the phase is energized, if the bar magnet is forced to rotate out of alignment, a torque will be applied to it by the magnetic field to bring it back into alignment. It is this magnetic pull (or push) between the rotor and the stator poles that creates the motor torque.

Motors often have more than two windings per phase. Figure 4.3 shows one phase made out of four windings that are connected in series for a four-pole motor.

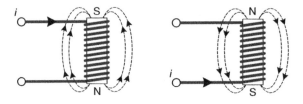

Figure 4.1   Magnetic polarity of an electromagnet can be changed by changing the current direction

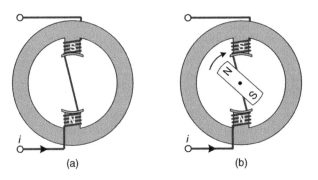

Figure 4.2   One-phase stator. (a) One phase winding made with two coils. (b) Bar magnet rotor

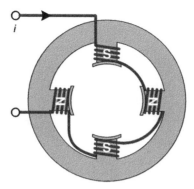

Figure 4.3   Winding of one phase for a four-pole motor

4.1.1   Electrical and Mechanical Cycles

Mechanical cycle is defined as 360 mechanical degrees (360° M) of rotor rotation. In one mechanical cycle, a point on the rotor goes through a complete rotation to end up exactly at the same location. One electrical cycle is defined as 360 electrical degrees (360° E). When one electrical cycle is completed, the *magnetic* orientation of the rotor becomes the same as its initial orientation.

Figure 4.4 shows a two- and a four-pole rotor. If the two-pole rotor is rotated clockwise 360° M, it will have exactly the same magnetic orientation. On the other hand, the four-pole rotor needs to be rotated only 180° M to put the rotor back in the same initial magnetic orientation. This relationship can be generalized by the following equation:

$$\theta_e = \frac{p}{2}\theta_m$$

where θ_e is the electrical position, θ_m is the mechanical position, and p is the number of poles. In the case of the two-pole rotor, as shown in Figure 4.4a, $p = 2$ and, therefore, $\theta_e = \theta_m$. In the case of the four-pole rotor, as shown in Figure 4.4b, $p = 4$; hence $\theta_e = 2\theta_m$. If $\theta_m = 180°$ M, then $\theta_e = 360°$ E (half a rotation of the rotor is equal to one electrical cycle).

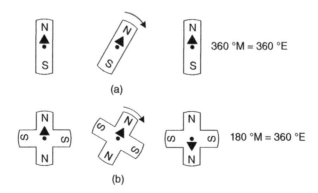

Figure 4.4   Electrical and mechanical cycles. (a) Two pole rotor. (b) Four pole rotor

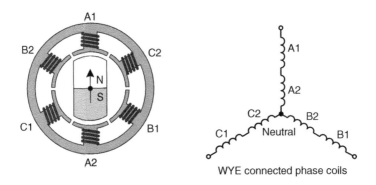

Figure 4.5   Three-phase stator and WYE-connected phases

4.1.2   Three-Phase Windings

In motion control applications, three-phase motors are very common. The concept shown in Figure 4.2a can be extended to a three-phase stator by placing three-phase windings 120° M apart from each other as shown in Figure 4.5. The resulting stator has three-phases with two-poles in each phase [25]. The two windings of each phase are connected in series. The three-phases are WYE connected. This is the most common type of connection, but the phases can also be DELTA connected. In WYE-connected motors, the common terminal (neutral) is often not accessible. There are only three lead wires coming out of the motor.

4.2   Rotating Magnetic Field

When a phase is energized, the rotor gets attracted to it and rotates toward it to align with the phase poles. To keep the rotor spinning, the next phase must be energized, while the current one is switched off. Essentially, as the rotor gets near alignment with the current phase, the stator magnetic field needs to be rotated to its next orientation so that the rotor remains in motion toward the next set of energized stator poles. Consequently, a rotating magnetic field must be generated by the stator to keep the rotor spinning [11].

4.2.1   Hall Sensors

We need to know the angular position of the rotor to create a rotating field. Typically, three Hall sensors are used for this purpose. They are usually mounted on a printed circuit board and attached to the inside of the enclosure cap at the back end of the motor. Most motors have a small set of magnets mounted at the end of the rotor shaft to trigger the Hall sensors. These magnets are aligned in the same way as the main magnets of the rotor itself. Every time one of these small magnets passes by a sensor, the sensor gives a low or high signal indicating the S or N pole of the magnet. As a result, we can determine the rotor's position. The Hall sensors are mounted either 60° or 120° apart from each other. Each sensor coincides with one of the magnetic phase circuits of the three-phase stator.

As can be seen in Figure 4.6, in every 60° electrical rotation of the rotor magnetic field, one of the Hall sensors changes its state. Therefore, the three Hall sensors divide the 360°

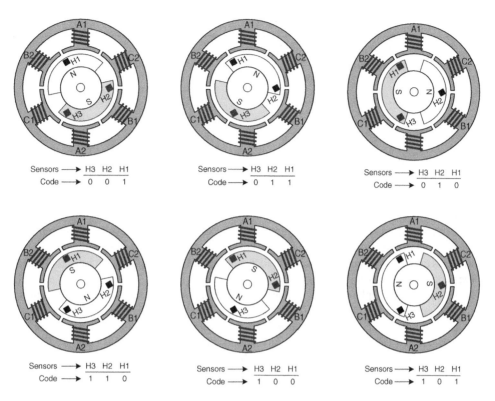

Figure 4.6 Three Hall sensors create six segments for the course measurement of the absolute position of the rotor field. Each segment is identified by a unique three-digit code formed by combining the signals from the sensors. Here it was assumed that the sensor gives a high signal (indicated by "1") when it is triggered by the North Pole of the rotor

region into six segments of 60° each. This approach provides a course measurement of the absolute position of the rotor field as it lands into one of these segments. Each of these segments corresponds to a three-digit code such as "001" or "010." Here, each digit is the high or low signal from one of the sensors.

4.2.2   Six-Step Commutation

Torque is produced from the interaction of the stator and rotor magnetic fields. For optimal torque production, the angle between these two fields needs to be maintained at 90°. This angle is called the *torque angle*. The 90° angle ensures that the rotor is pulled by the stator in the strongest possible way to create the maximum torque. *Brushed* DC motors maintain this optimal 90° angle using slip rings and brushes. In the case of a *brushless* motor, the phases must be switched ON/OFF electronically in a certain sequence to keep the two fields in proper orientation to produce torque. The electronic switching is called *commutation*.

The most common commutation algorithm is the six-step commutation. As explained in Section 4.2.1, the three Hall sensors divide the 360° electrical rotation region into six segments

Table 4.1   Six-step commutation for the brushless motor in Figure 4.7

Stator field	Hall sensor	Phase A voltage	Phase B voltage	Phase C voltage	A1	A2	B1	B2	C1	C2
0°	001		−	+			S	N	N	S
60°	011	+	−		N	S	S	N		
120°	010	+		−	N	S			S	N
180°	110		+	−			N	S	S	N
240°	100	−	+		S	N	N	S		
300°	101	−		+	S	N			N	S

of 60° each. As the rotor magnetic axis falls into one of these segments, a pair of stator windings must be energized to keep the stator magnetic field ahead of the rotor. The rotor will be attracted by the stator field and will turn toward it. When the rotor enters the next segment, another set of windings is energized to move the stator field ahead again. In this fashion, the rotor is "pulled" by the stator field and continues to turn.

Table 4.1 shows the phase voltages that need to be applied to create a rotating field in the motor. The stator field is advanced in 60° increments. The Hall sensor code indicates where the rotor field is at a given instant. For example, let us assume that the rotor is rotating clockwise (CW) and has just entered the segment indicated by the "001" Hall sensor code as shown in the top diagram of Figure 4.7. At this instant, the stator field is moved to horizontal position at 0° by the algorithm. The rotor field is along the North Pole direction of the rotor. Therefore, the instantaneous angle between the stator and the rotor fields is 120°. To orient the stator field in the horizontal direction, we need B1 and C2 to be South Poles, and C1 and B2 to be North Poles. This is accomplished by applying positive voltage to Phase C and negative voltage to Phase B as shown in Table 4.1. Note that in this sequence, one of the phases is always left open (floating). Keep in mind that the rotor is still moving CW.

While the stator field is held horizontal, the rotor is pulled by the stator field and rotates CW toward the stator field. When the rotor sweeps 60° and reaches the end of the "001" segment, the instantaneous angle between the stator and the rotor fields is down to 60° (still referring to the top diagram in Figure 4.7). As the rotor enters the next segment, indicated by the "011" Hall sensor code, the algorithm moves the stator field CW to 60° orientation (second diagram in Figure 4.7). The instantaneous angle between the fields becomes 120° again. The algorithm is now in its second step. The proper windings to be energized in this step are given in the second row of Table 4.1. At any given time, two phases of the motor are applied voltages and the third phase is floating.

In this scheme, the goal is to maintain the stator field 90° ahead of the rotor field to produce *optimal* torque. However, because the stator field is advanced by 60° increments based on the Hall sensor codes, the actual angle varies from 60° to 120° [2]. Hence, the desired 90° optimal angle cannot be maintained at all times, while the stator and rotor fields are rotating. Instead, an *average* torque angle of 90° is maintained, which leads to an undesirable fluctuation in torque called *torque ripple* [1].

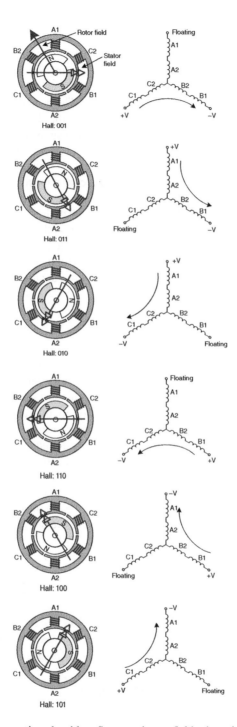

Figure 4.7   Six-step commutation algorithm. Stator and rotor field orientations and active phase windings in each step to rotate the rotor CW

4.3   AC Servo Motors

A typical AC servo motor consists of a three-phase stator with distributed windings, a rotor
with permanent magnets and a position sensor to determine the rotor position (Figure 4.8).
There are other names used for the same motor such as "permanent magnet synchronous motor
(PMSM)" and "permanent magnet AC motor (PMAC)." The primary distinguishing feature of
an AC servo motor from a brushless DC motor (BLDC) is the shape of the back electromotive
force (emf) voltages, which are trapezoidal in BLDCs and sinusoidal in the AC servo motors.

The six-step commutation given in Table 4.1 switches currents from phase to phase, which
leads to torque ripple. To improve motor operation, the stator windings of an AC servo motor
are supplied with *sinusoidal currents*, which create a continuously rotating magnetic field in
the motor. Unlike the six-step commutation, the currents are *apportioned* between the phases.
At any given instant, some amount of current flows in each of the windings. The magnitude and
the direction of each phase current are determined based on the instantaneous rotor position.
As explained in Section 4.2.1, Hall sensors divide the cycle into six 60° regions. For current
apportioning to work, we must know the rotor position with much higher accuracy. Hence, an
AC servo motor will have a high precision position sensor, such as an encoder, making these
motors more expensive.

4.3.1   Rotor

The rotor of an AC servo motor is cylindrical and constructed using solid or laminated iron core
and permanent magnets that are radially magnetized. Motors with 4-, 6- or 8-poles are common. Several magnet configurations are possible such as the surface-mounted and surface-inset
magnets as shown in Figure 4.9. By far the most common configuration is the surface-mounted
version where the magnets are bonded on the surface of the rotor. Due to the simple construction, these motors tend to be relatively inexpensive. Ferrite magnets are the common choice
as they are relatively inexpensive. Alloy magnets such as neodymium–ferrite–boron (NdFeB)
have higher flux density per unit volume [28]. Therefore, the motor size can be reduced and
higher torque can be produced but these magnets are more expensive.

In the surface-mounted design, the same torque can be obtained using smaller magnets in
comparison to the inset design. Also, the surface-mounted version provides the highest airgap

Figure 4.8   AC servo motors (Reproduced by permission of Emerson Industrial Automation [12])

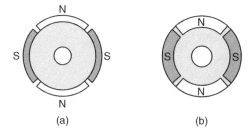

Figure 4.9 Four-pole rotor with (a) surface-mounted, and (b) inset-mounted magnets

flux since there are no other elements between the magnets and the stator poles. The relative permeability of the magnet material is almost equal to that of air. Therefore, the magnets act like extensions of the airgap leading to insignificant saliency and uniform flux density in the airgap. One disadvantage is the structural integrity of the rotor. In high speeds, the magnets can separate from the surface due to the centrifugal forces. In some motors, the rotor assembly is inserted inside a stainless steel cylindrical sleeve to hold the magnets in place. In other designs, the rotor assembly is wrapped using a tape. The surface-mounted design tends to result in lower inertia which is desirable for higher dynamic performance.

In the surface-inset design, the magnets are mounted inside grooves on the rotor. This provides a stronger mechanical integrity and a uniform rotor surface. Due to the difference between the relative permeability of the rotor iron and the magnets, this design has saliency. When a rotor magnet is aligned with a stator pole, a certain stator inductance is obtained. When the rotor rotates and the rotor iron between the magnets aligns with a stator pole, a different inductance is obtained. Such inductance variations are negligible in the surface-mounted designs as the permeability of air and magnets is almost the same. However, due to the stronger construction, the inset rotors enable much higher speeds.

4.3.2 Stator

The stator is made of laminations, which are thin sheets of metal. The laminations are punched out of sheet metal, stacked, and welded together on the side to form the stator. The laminations reduce the energy losses due to Eddy currents compared to a solid core. Slot insulation materials and phase winding coils are inserted and the coil ends are tied. Heat hardened varnish on the windings finishes the assembly.

Stator windings are designed to obtain *sinusoidal back emf* voltages. Therefore, AC servo motors are sometimes called "sinusoidal motors." The sinusoidal phase currents and back emf voltages result in no torque ripple. The motor operates smoothly even in very low speeds.

The stator of an AC induction motor is made of a stack of thin metal sheets.

4.3.2.1 Distributed Windings

Stator windings in Section 4.1 were constructed by winding coils on salient poles. Most motors use a different construction called *distributed windings*. In this case, the stator consists of slots between teeth as shown in Figure 4.10a. Large coils are prepared and inserted into the stator

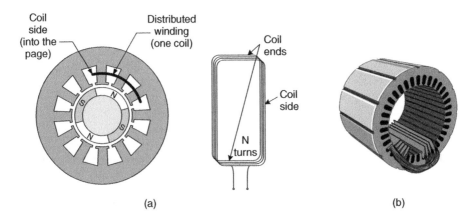

Figure 4.10   Stator for distributed windings. (a) Stator with slots and a coil for phase winding. (b) Phase windings in stator slots (Reproduced by permission of Siemens Industry, Inc.) [25]

slots by placing the coil sides in the slots as shown in Figure 4.10b. All coil ends are then grouped and shaped to be out of the way of central opening of the stator so that the rotor can later be inserted.

Figure 4.11 shows the distributed phase A windings for a four-pole motor. It consists of four coils inserted into the slots. The three stator teeth in each coil form a stator pole similar to the salient poles in Figure 4.2. The current direction in a coil defines the polarity of the coil. Therefore, depending on the current direction in the coil, the teeth in the coil create a North or South Pole [13, 14]. All coils of phase A are connected in series, but they are inserted into the slots with alternating coil directions. For example, if the first coil is wound in the clockwise direction on the teeth, the next coil is inserted so that it is wound in the counter-clockwise direction on the teeth. This way, when the current is supplied to the phase, the first coil creates a North Pole, the second coil creates a South Pole, the third coil creates a North Pole, and the fourth one creates a South Pole. Phases B and C are wound similarly and placed in the appropriate slots in the stator to form the complete three-phase winding.

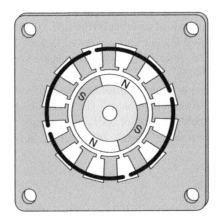

Figure 4.11   Distributed phase A winding for a three-phase four-pole motor

4.3.2.2   Coil Pitch

A coil pitch (or span) is defined as the angular distance between the two sides of a coil [19, 24, 26]. A *pole pitch* is defined as the angular distance between the rotor pole centers (Figure 4.12). If the coil pitch is equal to the pole pitch, then the winding is called *full pitch winding*. If the coil pitch is shorter than the pole pitch, the winding is called *fractional-pitch winding* [22]. These windings are also referred to as short-pitched or chorded windings.

4.3.2.3   Flux Linkage and Back emf Voltage with Full-Pitch Coils

Figure 4.13 shows a four-pole motor with full-pitch coils. As the rotor rotates, the flux linkage λ (amount of rotor magnet flux passing through the coil) will vary [14, 16]. At position 1, the flux linkage is positive (out of the N pole of the rotor). As the rotor rotates clockwise toward position 2, the portion of the N pole facing the coil reduces, while more and more of

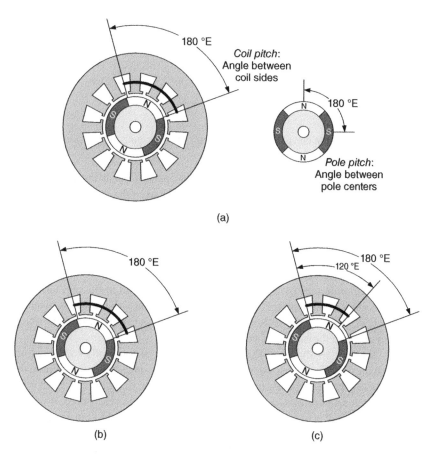

Figure 4.12   Winding terminology. (a) Coil pitch and pole pitch. (b) Full-pitch winding. Coil pitch is equal to pole pitch. (c) Fractional-pitch winding. Coil pitch is shorter than pole pitch

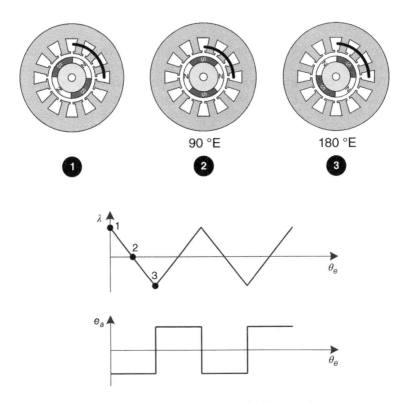

Figure 4.13   Flux linkage in one phase coil as the rotor of a four-pole full-coil pitch motor rotates

the S pole faces the coil. Therefore, the flux linkage drops in a linear fashion. At position 2, equal portions of the S and N poles face the coil, which makes the flux linkage zero. As the rotor continues to rotate, the flux linkage continues to drop due to the increased exposure to the S pole. At position 3, the entire S pole faces the coil. Hence, the maximum amount of negative flux linkage is reached. After this point, as the rotor continues to rotate in the same direction, the flux linkage increases linearly since more of the N pole faces the coil. When the rotor completes a full rotation, the cycle is repeated twice.

When a rotor magnet passes by the coil, a voltage e is induced at the coil terminals. This voltage is called the *back emf* voltage. It opposes the coil input voltage and is given by Faraday's law $e = d\lambda/dt$, where λ is the flux linkage [7]. Then, the slope of the flux linkage plot in Figure 4.13 is equal to the back emf voltage e_A.

The phase A winding, as shown in Figure 4.11, consists of four coils connected in series. Each coil contributes to the overall back emf voltage of the phase winding. Since these are all full-pitch coils, the back emf waveforms of all coils will be in phase with each other. As a result, when the back emf voltage of the entire phase A winding is measured, it will have the same shape as e_A, as shown in Figure 4.13, but the amplitude will be four times as much as the amplitude of the waveform of a single coil.

In the case of a three-phase motor, the back emf in each phase has the same shape as shown in Figure 4.13. However, due to the phase-winding arrangements, the back emf voltages are shifted by 120° E as shown in Figure 4.14. Furthermore, in real motors, the ideal square back

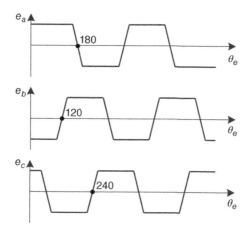

Figure 4.14   Back emf voltage in each phase winding of a three-phase stator

emf waveform turns into a trapezoidal waveform due to the flux leakages between the adjacent magnets.

4.3.2.4   Sinusoidal Back emf

Motor designers can shape the back emf voltages by using design features such as fractional-pitch coils, fractional-pitch magnets, and fractional-slot stators.

The number of slots per pole per phase q is found from [24]

$$q = \frac{S}{mp}$$

where S is the number of slots, p is the number of poles, and m is the number of phases. If q is an integer number, the motor is called an *integral slot motor*. Otherwise, it is called a *fractional slot motor*. For example, the motor shown in Figure 4.11 has 12 slots, 4 poles, and 3-phases. Since $q = 12/(3 \cdot 4) = 1$, it is an integral slot motor, whereas, if the motor had 18 slots, it would be a fractional slot motor since $q = 18/(3 \cdot 4) = 1.5$ is not an integer.

If the motor is an integral slot motor, the back emf waveforms of the coils in a phase winding will all be in phase. On the other hand, in fractional slot motors, the back emf waveforms of the coils are shifted in phase (Figure 4.15). As a result, when the coils of a phase are connected in series, the shape of the overall back emf waveform measured at the terminals of the phase winding will have a different waveform shape than the shape of each coil's back emf waveform. Furthermore, due to flux leakages, the corners of the waveform will be rounded. By carefully selecting the parameters such as the number of slots, coil pitch, pole pitch, and, fractional magnets the back emf voltages of an AC servo motor can be shaped to be sinusoidal. For more information, the reader is referred to the excellent reference [14].

4.3.3   Sinusoidal Commutation

In sinusoidal commutation, each phase of the stator is supplied with a sinusoidal current (Figure 4.16a). Each phase winding then generates a magnetic field along its magnetic axis.

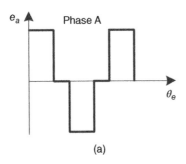 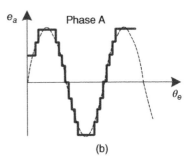

(a)   (b)

Figure 4.15   Back emf waveforms in one phase winding of an integral slot motor and a fractional slot motor. (a) Integral slot motor. All coil back emf voltages are in phase with each other. Hence, the overall phase winding back emf has the same shape as coil voltages but the amplitude is the sum of all coil voltages. (b) Fractional slot motor. Individual coil back emf voltages are shifted in phase. The overall phase winding back emf voltage is approximately sinusoidal

The field pulsates as its magnitude varies and its direction changes back and forth along the magnetic axis of the winding based on the sinusoidal phase current. Figure 4.16b shows the pulsating field F in phase A of the stator as the phase current changes sinusoidally. In this example, one side of the phase A winding is in slot A1, while the other side is in slot A2. In slot A1, the current comes out of the page, while in slot A2, it goes into the page. If you wrap your fingers from A1 to A2, your thumb will be in the direction of the *magnetic axis* for phase A.

When pulsating fields of each phase are combined, they result in a rotating magnetic field [24]. For example, at *position 1*, as shown in Figure 4.17, the phase A current is at its maximum value and positive (I_{max}), while the phase B and C currents are at half this value and are negative ($i_a = i_b = -\frac{1}{2}I_{max}$). Each of these phase currents generates a corresponding magnetomotive force (MMF) F.

The vector sum of these forces gives the total magnetic field generated by all three phases

$$\vec{F}_R = \vec{F}_a + \vec{F}_b + \vec{F}_c$$

At *position 1*, the vertical components of $\vec{F}_b$ and $\vec{F}_c$ cancel out. Therefore, the resultant magnetomotive force is just given by the horizontal components as

$$\vec{F}_R = \left[F_{max} + 2\frac{F_{max}}{2} \cos(60) \right] \hat{i} = [1.5\,F_{max}]\hat{i}$$

where $\frac{F_{max}}{2} \cos(60)$ is the magnetomotive force generated by phases B and C. The magnetomotive force from phase A is at its maximum F_{max}. The resultant vector is along the positive direction of the phase A magnetic axis and has a magnitude of $1.5F_{max}$.

At *position 2*, the phase C current reaches its maximum value but is in the negative direction ($i_c = -I_{max}$). Phases A and B currents are positive and equal to half of i_c. Then, the magnetomotive force vectors are

$$\vec{F}_a = 0.5F_{max}\,\hat{i}$$
$$\vec{F}_b = 0.5F_{max}\,[-\cos(60)\hat{i} - \sin(60)\hat{j}]$$
$$\vec{F}_c = F_{max}\,[\cos(60)\hat{i} - \sin(60)\hat{j}]$$

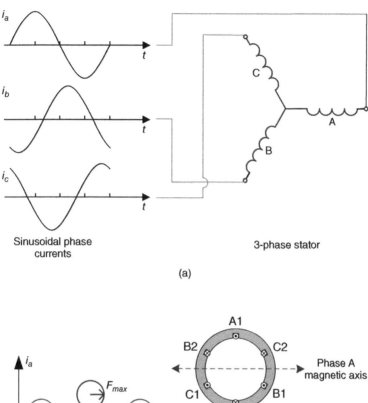

Sinusoidal phase
currents

3-phase stator

(a)

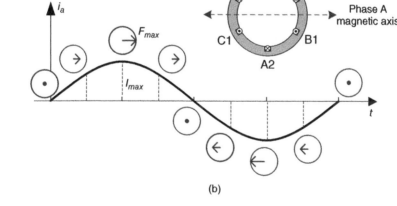

(b)

Figure 4.16   Sinusoidal phase currents and pulsating magnetic field in phase A. (a) Sinusoidal phase currents shifted by 120. (b) Pulsating magnetic field in phase A. Its magnitude varies and its direction changes back and forth along the magnetic axis of the winding

where $\hat{i}$ and $\hat{j}$ are the unit vectors in the x and y directions of the coordinate frame, respectively. Then, the resultant magnetomotive force can be found from the vector sum

$$\vec{F}_R = 1.5F_{\max}[\cos(60)\hat{i} - \sin(60)\hat{j}]$$

This resultant vector is along the negative direction of the phase C magnetic axis with a magnitude of $1.5F_{\max}$.

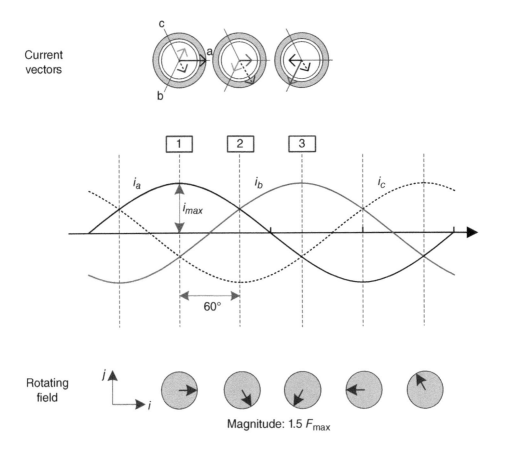

Current vectors

Rotating field

Magnitude: 1.5 F_{max}

Figure 4.17   Rotating magnetic field as a result of sinusoidal phase currents

At *position 3*, the phase B current reaches its maximum value and is positive ($i_b = I_{max}$). Phases A and C currents are negative and half of i_b. Then, the resulting total magnetomotive force is

$$\vec{F}_a = -0.5 F_{max}\, \hat{i}$$
$$\vec{F}_b = F_{max}\, [-\cos(60)\hat{i} - \sin(60)\hat{j}]$$
$$\vec{F}_c = -0.5 F_{max}\, [\cos(60)\hat{i} - \sin(60)\hat{j}]$$

The resultant magnetomotive force can be found from the vector sum

$$\vec{F}_R = 1.5 F_{max}[-\cos(60)\hat{i} - \sin(60)\hat{j}]$$

This resultant vector is along the positive direction of the phase B magnetic axis and has a magnitude of $1.5 F_{max}$. From these observations, we see that when sinusoidal currents are supplied to the phases, a constant amplitude resultant magnetomotive force is produced that rotates.

The speed at which the magnetic field rotates is called the *synchronous speed* n_s and is given by

$$n_s = \frac{120 \cdot f}{p} \tag{4.1}$$

where f is the frequency of the phase currents in Hertz and p is the number of poles.

4.3.4   Torque Generation with Sinusoidal Commutation

AC servo motors with sinusoidally commutated drives have no torque ripple since the sinusoidal phase currents result in a smooth rotating field. In the sinusoidal commutation scheme, the amplifier varies the current for each phase sinusoidally based on the instantaneous rotor position [2]. Hence, there is a requirement to measure the rotor position with much higher accuracy than the six positions obtained from the Hall-effect sensors. Typically, an encoder is used for this purpose making the cost of implementation much higher. The motor phase currents as a function of rotor position are

$$I_a = U_c \sin(\theta)$$
$$I_b = U_c \sin\left(\theta + \frac{2\pi}{3}\right) \tag{4.2}$$
$$I_c = U_c \sin\left(\theta + \frac{4\pi}{3}\right)$$

where U_c is the current command signal to the amplifier and θ is the rotor position.

Assuming no losses, equating mechanical, and electrical power per phase gives

$$T\omega_m = e(\theta)i(\theta),$$

where ω_m is the rotor mechanical speed, T is the torque, $e(\theta)$ and $i(\theta)$ are the phase back emf voltage and current, respectively. The equation can be rearranged as

$$T = \frac{e(\theta)}{\omega_m}i(\theta)$$
$$= k_e(\theta)i(\theta)$$

Since the back emf voltages are sinusoidal, $k_e(\theta) = K_e \sin(\theta)$, where K_e is the voltage constant for the motor. Then, the torque equations for a three-phase stator are

$$T_a = I_a K_e \sin(\theta)$$
$$T_b = I_b K_e \sin\left(\theta + \frac{2\pi}{3}\right) \tag{4.3}$$
$$T_c = I_c K_e \sin\left(\theta + \frac{4\pi}{3}\right)$$

The total torque generated by the motor is the sum given by

$$T = T_a + T_b + T_c \tag{4.4}$$

If we substitute Equations (4.3) and (4.2) into Equation (4.4),

$$T = [(U_c \sin(\theta))K_e \sin(\theta)] + \left[\left(U_c \sin\left(\theta + \frac{2\pi}{3}\right)\right)K_e \sin\left(\theta + \frac{2\pi}{3}\right)\right]$$
$$+ \left[\left(U_c \sin\left(\theta + \frac{4\pi}{3}\right)\right)K_e \sin\left(\theta + \frac{4\pi}{3}\right)\right]$$

Rearranging gives

$$T = K_e U_c \left[\sin^2(\theta) + \sin^2\left(\theta + \frac{2\pi}{3}\right) + \sin^2\left(\theta + \frac{4\pi}{3}\right)\right]$$

After simplifications,

$$T = 1.5 K_e U_c$$

It is interesting to note that the torque is no longer a function of the rotor angle θ and is constant. It is also easier to produce sinusoidal currents compared to the switching square current pulses required in the case of the six-step drives. Therefore, sinusoidally commutated drives eliminate torque ripple when combined with AC servo motors.

4.3.5   Six-Step Commutation of AC Servo Motors

The six-step commutation explained in Section 4.2.2 simplifies motor feedback requirements since it only uses simple Hall sensors for the determination of the rotor position. Also, current control and switching can be done using simple analog circuitry leading to an inexpensive system implementation. Therefore, sometimes an AC servo motor may be coupled to a drive with six-step commutation.

In six-step commutation, current flows into a phase and flows out of another phase, while the third phase is not connected. Assuming no losses in the motor and using conservation of energy

$$e_{ph-ph} \cdot I_{ph} = T\omega_m \tag{4.5}$$

where e_{ph-ph} is the phase-to-phase back emf voltage. If the motor speed ω_m is constant, the shape of the torque profile T will be the same as the phase-to-phase back emf signal of the active phase since in six-step commutation, the active phase current I_{ph} is constant.

Figure 4.18 shows the torque generated by an AC servo motor driven by six-step commutation. In each $60°$ segment, Hall sensors determine which phase-to-phase back emf voltage is active in torque generation. For example, in the $60–120°$ segment, the e_{a-b} signal is active. Therefore, the arc of that signal is the torque curve in that segment.

Looking at the $60–120°$ segment of the torque curve, we can calculate the theoretical torque ripple as

$$T_{min} = T_p \sin(60)$$
$$T_{max} = T_p \sin(90)$$
$$\text{Torque ripple} = \frac{T_{max} - T_{min}}{T_{max}} \approx 13\% \tag{4.6}$$

where T_p is the peak torque.

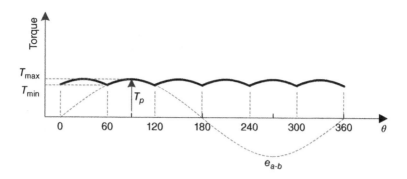

Figure 4.18   Torque generated when an AC servo motor is coupled to a six-step drive

Six-step commutation of an AC servo motor results in *torque ripple* (or pulsation) [5]. This is mainly due to switching phase currents every 60° E. The resulting magnetic field jumps as opposed to the smooth rotation obtained with sinusoidal phase currents. Torque ripple is most noticeable at low speeds. Usually, it can be observed in the form of speed fluctuations. This speed ripple also depends on the dynamic characteristics of the machine and the load. A highly compliant coupling or load will suffer more from the torque ripple and show oscillatory behavior. Similarly, a system with low inertia will exhibit the same behavior. Therefore, in positioning applications, six-step commutation is not used.

4.3.6   Motor Phasing with Encoders and Hall Sensors

All brushless motors require some type of a phase search when they are powered on to establish a relationship between the zero position of the motor commutation cycle and the zero position of the feedback device (usually incremental encoder).

The goal is to "mark" one of the phase positions as the beginning position for the incremental encoder counts. This way, the encoder can be used to precisely know the angular position of the rotor with respect to the phase windings of the motor. This information is then used by the commutation algorithm of the motion controller to apply the sinusoidal phase currents correctly at any instant of the rotor motion.

The phasing search causes small movements of the motor shaft. The more poles the motor has, the smaller the shaft movement will be. There are various phasing algorithms developed by motion controller manufacturers [3]. The most basic one involves applying phase currents to two of the three-phases of the motor to force the motor to rotate to a known position in the phase cycle and to mark that position as the encoder zero. First, the controller opens the feedback loop of the motor so that it can be spun freely. Then, it applies current to enter phase A and exit phase B. This causes the motor to rotate through a small angle until the rotor magnets align with winding poles. Once the rotor is locked into position, Hall sensors are read, and the motion controller resets the encoder counter register of the motor to zero to mark this position as the beginning position. Finally, the position feedback loop is closed to get ready to commutate the motor with sinusoidal phase currents.

The advantage of this algorithm is that it is more robust for systems with high friction and/or large directional loads such as gravity. On the other hand, it may cause too much movement for certain applications. Another phasing method makes two guesses for the phasing reference point. It applies a torque command at each guess and observes the response of the motor in each case. Based on the magnitude and direction of the two responses, the motion controller calculates the phasing reference point to reset the encoder counter to zero. This type of search is gentler as it causes less movement of the shaft and is effective for most systems. However, it may not be too reliable in systems with significant friction or directional load.

Certain applications cannot tolerate any movement of the motors on power up. In these applications, motors with absolute position feedback sensors must be used along with controllers that are capable of utilizing such sensors in power-up sequences.

4.4 AC Induction Motors

Three-phase AC induction motors (Figure 4.19) are extensively used in industrial machinery. The main advantages of these motors include simple and rugged construction, easy maintenance, and economical operation. For a similar output rating, an AC induction motor is smaller in weight and has much less rotor inertia compared to brushed motors. One significant disadvantage of the induction motor is that torque control is complicated due to the rotor field induced by the stator field. Until recently, induction motors were mainly used in constant speed applications such as in conveyors. But with the advances in power electronics and computing technologies, induction motors (inverter- or vector-duty) are now widely used in variable speed and motion control applications.

4.4.1 Stator

Like the stator of an AC servo motor, the stator of an AC induction motor is also made of a stack of thin metal sheets. Slot insulation materials are inserted into the stator slots. This is followed by the insertion of the phase-winding coils based on a certain pattern (Figure 4.20a) [25]. Phase-to-phase insulation materials are placed between the coils of each phase winding and the wiring leads are connected and soldered. The end turns of each coil is then tied which keeps the coils from vibrating while the motor is spinning. Finally, the stator assembly is dipped into a varnish and baked in an oven to harden it (Figure 4.20b).

Figure 4.19   Vector-duty AC induction motor (Reproduced by permission of Marathon™ Motors, A Regal Brand) [18]

(a)   (b)

Figure 4.20 AC induction motor stator. (a) Stator windings in slots (Reproduced by permission of Siemens Industry, Inc.) [25]. (b) Completed stator (Reproduced by permission of Siemens Industry, Inc.) [25]

Most induction motors have sinusoidal distributed phase windings similar to those explained in Section 4.3.2. The coils of the winding of a phase are inserted into slots around the stator. When energized, each phase winding can generate a certain number of motor poles. Typical numbers of poles are 2, 4, 6, or 8.

An induction motor operates at its rated fixed speed when supplied with the voltage and frequency specified on its nameplate. However, some motors have external connections that can be altered to change the number of poles in the stator. Therefore, they can run at two speeds and are called *consequent pole* motors [10, 21, 27]. The low speed is always half of the high speed. For example, 2/4 or 4/8 pole combinations are common. The external connections arrange the stator windings in series-delta or parallel-wye configurations. The series delta results in low speed and the parallel-wye in high speed.

The stator windings are covered with insulation for protection against high temperatures. The National Electrical Manufacturers Association (NEMA) established four insulation classes: A, B, F, and H according to the maximum allowable operating temperature of the winding. Class A is for temperatures up to 105 °C, class B is for up to 130 °C, class F is for up to 155 °C, and class H is up to 180 °C. Class F is the most commonly used type. These temperature limits include a standardized ambient temperature of 40 °C and an additional 5–15 °C of margin at the center of the windings called the motor's hot spot [4, 6, 9].

In addition to these four classes of motors, *inverter duty* or *vector duty* class describes AC induction motors that are specifically designed for use in variable speed applications and with motion control drives. The high-frequency switching, voltage spikes, and extended periods of motor operations at low speed lead to higher temperatures in the windings. The inverter duty motors have high-temperature insulating materials that can withstand higher operating temperatures than the regular AC induction motors.

4.4.2 Rotor

The most common induction motors have a squirrel cage rotor as shown in Figure 4.21. It is constructed using heavy-gauge copper or aluminum bars that are short-circuited by end caps. The construction resembles a hamster-wheel or a squirrel cage. The conductor bars are inserted

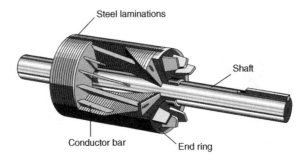

Figure 4.21   Squirrel cage rotor of an AC induction motor (Reproduced by permission of Siemens Industry, Inc. [25])

in a stack of steel laminations. The steel core helps concentrate the flux lines between the rotor conductors and reduces air reluctance. The slight skew of the bars helps reduce a humming sound. The rotor and the shaft are rigidly connected.

4.4.3   Motor Operation

The stator magnetic field induces currents in the conductors of the rotor. This current flows through the rotor bars and around the end rings. As a result, a magnetic field is generated around each of the rotor conductor bars (Figure 4.22). The cage becomes an electromagnet with alternating N and S Poles due to the rotating field of the stator. The magnetic field of the rotor tries to keep up with the magnetic field of the stator. Their interaction generates torque and runs the motor.

The rotating magnetic field of the stator induces voltages in the rotor bars. Based on Faraday's induction law, the induced voltage is proportional to the time rate of change in the flux lines that cut the rotor bars. If the rotor spins at the same speed as the stator field, there will be no relative motion between the rotor and the stator field. As a result, no induced voltage (or torque) will be generated and the motor would not run. The rotor, therefore, runs at a slower speed than the stator. The difference between the rotor and stator speeds is called *slip*.

As explained in Section 4.3.3, the speed of the stator magnetic field is called *synchronous speed* (n_s) and is given by Equation (4.1). The actual speed of the rotor (n) will be slightly less

Figure 4.22   Induced rotor currents and resulting magnetic fields (Reproduced by permission of Siemens Industry, Inc. [25])

than the synchronous speed. The nameplate of induction motors provides the actual speed. Slip is expressed using the following equation [15]:

$$s = \frac{n_s - n}{n_s} \tag{4.7}$$

When the motor starts, the magnetic field created by the stator windings is immediately at the synchronous speed, while the rotor is stationary. Therefore, at the start, the slip is $s = 1.0$ (100%). The motor draws a high inrush current during start-up. This current can be up to 6 times the full-load current listed on the nameplate and is also called the *locked-rotor current*. As the motor gradually gains speed, the motor current decreases. The rotor speed gets to a point where the torque demanded by the load equals the torque produced by the motor. At this point, the motor speed remains constant as long as the loading does not change. The motor speed must be less than the synchronous speed to produce useful torque and changes with the loading. Typically, the slip is in the range of 0.02–0.03 (2–3%) for a three-phase 60 Hz motor. Because the motor does not run at the synchronous speed, induction motors are also called *asynchronous* motors.

4.4.4   Constant Speed Operation Directly Across-the-Line

An AC induction motor can be directly powered from the three-phase power lines using a motor starter (Section 5.4) or it can be controlled by an AC drive. When operated directly across-the-line, the motor runs at a fixed speed and provides constant torque. On the other hand, if the motor is controlled by a vector drive, variable speed and torque are obtained as explained in Chapter 3, Section 3.5.2.

Three-phase AC induction motors are classified by NEMA into "A", "B", "C", and "D" classifications. These classifications are based on certain operating characteristics specified by NEMA [17]. When voltage is initially applied, the rotating magnetic field of the stator starts spinning instantaneously at the synchronous speed. Meanwhile, the rotor has to develop enough torque to overcome friction and accelerate its own and load inertia to reach the operational speed. The change in the torque as the rotor accelerates toward the synchronous speed is often shown as the torque–speed curve in Figure 4.23. This curve is for a NEMA B motor as this type of motor is the most commonly used one. The starting torque of a NEMA B motor using across-the-line starting typically produces 1.5 times its full-load running torque when the motor terminal voltage is the nameplate voltage.

Figure 4.23 also shows various operating points for a typical NEMA B motor. When the motor is first powered up, the stator magnetic field instantaneously reaches the synchronous speed. Initially, the rotor is stationary, which causes the NEMA B motor to develop 150% of its full-load torque. This is called the *starting torque*. As the motor accelerates, the torque drops slightly. The motor continues to speed up and the torque increases up to the *breakdown torque*. At this point, the torque for a NEMA B motor is about 200% of its full-load torque. As the speed increases past this point, the torque drops rapidly until the *full-load torque* is reached. Note that at the full-load torque, the motor is operating with a small slip.

As the NEMA B motor develops its starting torque, the starting current is about 600% of the full-load current. The starting current is also called the *locked rotor current* since the rotor is initially at rest. The *full-load current* is the stator current drawn from the line power while the motor is spinning with the applied rated voltage and frequency.

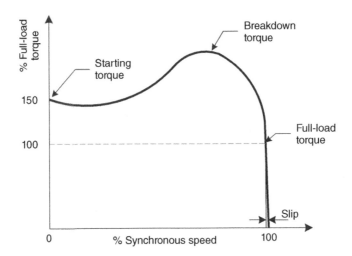

Figure 4.23 Torque–speed curve of a NEMA B type AC induction motor and various operating points [25]

The nameplate of the motor provides all the necessary information for proper application of the motor. A *service factor* is specified to determine the maximum horsepower at which the motor can be operated. If the service factor is 1.0, then the maximum horsepower is what is specified on the nameplate. However, a service factor higher than 1.0 shows that the motor can exceed the specified nameplate horsepower for brief periods of time. For example, if the service factor is 1.15 on a 2 HP motor, then the motor can be operated at $2 \times 1.15 = 2.3$ HP. But this operating condition should be used briefly; otherwise, the motor life will be reduced.

Using Equation (4.1) with $p = 4$ and $f = 60$ Hz, we can calculate the synchronous speed as 1800 rpm. The *actual* speed of the motor is most likely specified on its nameplate as 1775 rpm. This is called the *base speed* and is the actual speed at which the rotor will spin when connected to the line voltage and frequency specified on the nameplate. Using Equation (4.7), we can calculate the slip as 0.014 (1.4%).

4.4.4.1 Other NEMA Motors

As mentioned earlier, three-phase AC induction motors are classified by NEMA into "A", "B", "C", and "D" classifications. The difference is due to the changes in the rotor construction. Different rotor constructions produce different torque, speed, slip, and current characteristics as shown in Figure 4.24.

NEMA A motors have a torque speed curve similar to the NEMA B motor but with higher starting currents.

NEMA B motors have low rotor resistance, which produces low slip and high efficiency. They are used in applications where the motor needs to operate with a constant speed and load and without frequent turning on or off. Centrifugal pumps, fans, blowers, and small conveyors are such applications.

NEMA C motors have the same slip and full-load torque characteristics as the NEMA B motor but they have a higher starting torque (about 240% of full-load torque). In spite of the

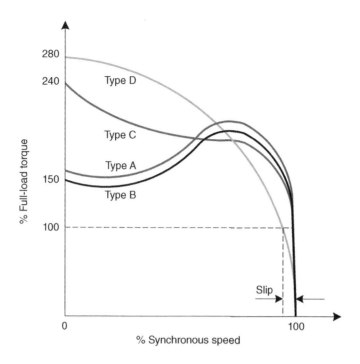

Figure 4.24   Torque–speed characteristics of NEMA A, B, C, and D motors

higher starting torque, they have a relatively low starting current. These motors are used with loads that are difficult to start such as crushers, big conveyors, and compressors.

NEMA D motors have high resistance rotors. The efficiency is decreased and the slip is increased typically to the range of 0.05–0.13 (5–13%). Their starting torque is about 280% of the full-load torque. Hence, these motors are used in applications that are very hard to start and require frequent starting, stopping, and reversing. Typical applications include cranes, punch presses, and hoists. The type D motor is the most expensive among all.

4.4.5   Variable Speed Operation with a VFD

The AC induction motor is a constant speed motor when operated across-the-line power. However, in many applications, we need to be able to change the speed of the motor. A popular method of controlling the speed of an AC induction motor is to vary the frequency of the input voltage supply to the motor. AC drives called *variable frequency drives (VFD)* are used for this purpose.

The air gap flux in the motor is proportional to the ratio of the motor voltage to frequency. When an AC induction motor is running at a constant speed, the flux remains constant since the voltage and frequency are constant. The torque is proportional to the flux. If the frequency is varied, the speed will vary, and the torque will also vary. To keep the torque constant while varying the speed, the voltage-to-frequency ratio must be kept constant. This ratio is known as the *volts per hertz (V/Hz) ratio*. Variable frequency drives take fixed voltage and frequency

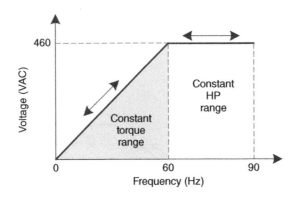

Figure 4.25 Constant torque and constant horsepower operating ranges for an AC induction motor controlled by a variable frequency drive (VFD) (Adapted by permission of Siemens Industry, Inc. [25])

AC supply and convert it into variable frequency and voltage supply. They implement control algorithms to keep the V/Hz ratio constant while varying the speed of the motor. Since the VFD controls the frequency and magnitude of the voltage, this control method is also called *scalar control*.

A VFD can operate the motor with constant flux from a few Hz to the motor's rated frequency (60 Hz in the United States). Because the flux remains constant, the motor operates with a constant torque. Therefore, this range of operation is called the *constant torque range*. For example, for a standard 460 VAC motor operating at 60 Hz, the ratio is: $V/Hz = \frac{460}{60} = 7.67$. Typically, at 60 Hz, the motor operates at its base speed of 1775 rpm as specified on its nameplate. To run this motor at half of the base speed (887 rpm), the VFD will reduce the motor voltage to 230 VAC and frequency to 30 Hz, so that the V/Hz ratio remains as 7.67.

In some applications, the motor must be operated at a faster speed than its base speed. This range of operation is called the *constant horsepower range*. The applied voltage remains fixed, while the speed (frequency) is increased. Therefore, the torque must decrease to keep the power constant. Figure 4.25 shows both of these operating ranges.

A VFD can start the motor at a reduced voltage and frequency. This prevents the high starting current (600% in the case of a NEMA B motor) as the motor is accelerated to the operating speed. The drive increases the voltage and frequency carefully while keeping the V/Hz ratio constant. VFDs can operate in open-loop or closed-loop mode. In industrial applications, the open-loop type is used most commonly. Open-loop VFDs have no feedback from the load. Therefore, they cannot self-regulate based on changes in the load condition. They can provide good speed regulation but lack fast dynamic performance to react to sudden changes in the load. VFDs are mostly used in regulating the speed of fans, pumps, mixers, and conveyors.

4.5 Mathematical Models

A motor converts electrical power into mechanical motion. Since the motor is an electromechanical device, its functionality can be described by the interaction of electrical and mechanical components as shown in Figure 4.26b.

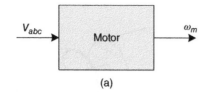

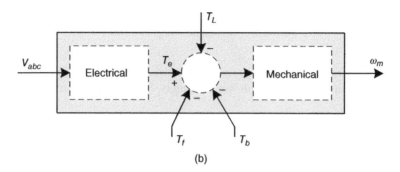

Figure 4.26 Motor conceptual model. (a) Motor converts voltage input into rotor speed. (b) Motor model includes electrical and mechanical components

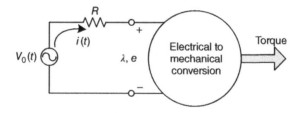

Figure 4.27 Per-phase circuit model for a motor

We can begin to model the electrical side of the motor by considering the per-phase electrical circuit shown in Figure 4.27. The motor's magnetic structure with windings generates a back emf voltage "e" at the terminals of the phase. This voltage opposes the input voltage and is given by Faraday's law $e = d\lambda/dt$ [7].

The circuit can be expressed by the following equation:

$$V_0(t) = iR + \frac{d\lambda}{dt} \tag{4.8}$$

The mechanical side can be modeled, as shown in Figure 4.28. Here, T_e is the torque generated by the electrical side of the motor. T_f models static friction, such as Coulomb friction, while T_b can account for any energy dissipation due to friction in the bearings of the motor. The torque generated by the electrical side works against the torques due to the static friction, viscous friction, and the load. The difference of these torques is used to accelerate the rotor to a mechanical rotational speed of ω_m. Using Newton's second law, we can write

$$T_e - T_L - T_f - T_b = J\dot{\omega}_m$$

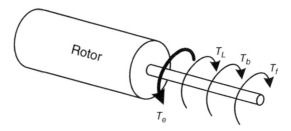

Figure 4.28   Mechanical model of a motor

where J is the rotor inertia. The viscous friction torque is often modeled as $T_f = b\,\omega_m$, where b is the viscous friction coefficient with Nm s/rad units. Substituting and rearranging give

$$T_e - T_L - T_f = J\dot{\omega}_m + b\,\omega_m \tag{4.9}$$

Also,

$$\omega_m = \frac{d\theta_m}{dt}$$

where θ_m is the rotor angular position (rad). Taking Laplace transform of both sides of Equation (4.9) with zero initial conditions yields

$$\omega_m(s) = \frac{1}{Js + b}\,(T_e - T_L - T_f) \tag{4.10}$$

The mechanical model in Equation (4.10) was implemented using Simulink® as shown in Figure 4.29a. This is a top-level block. When this block is opened, its details are revealed as shown in Figure 4.29b. The block takes inputs Te and TL as the torque generated by the electrical side and the load torque, respectively. It computes the angular speed wm and position theta. The signum function, Sign, was used to make sure that the static friction torque always opposes the input torque T_e. The J, B, and Tf values need to be specified for the simulated motor by entering them into MATLAB®.

4.5.1   AC Servo Motor Model

An AC servo motor receives phase voltages V_a, V_b, and V_c, which are then converted into torque by the stator and rotor field interactions. Starting from Equation (4.8), we can write the following equations for all phases of the motor:

$$V_a = i_a R + \frac{d\lambda_a}{dt}$$

$$V_b = i_b R + \frac{d\lambda_b}{dt} \tag{4.11}$$

$$V_c = i_c R + \frac{d\lambda_c}{dt}$$

Here, the flux linkage λ is defined as $\lambda = Li$, where L is the inductance and i is the current. The challenge is that the inductance is dependent on the rotor position and is not constant.

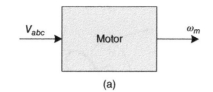

(a)

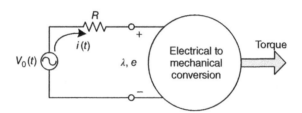

(b)

Figure 4.26 Motor conceptual model. (a) Motor converts voltage input into rotor speed. (b) Motor model includes electrical and mechanical components

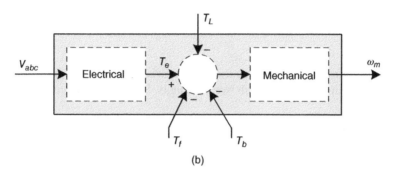

Figure 4.27 Per-phase circuit model for a motor

We can begin to model the electrical side of the motor by considering the per-phase electrical circuit shown in Figure 4.27. The motor's magnetic structure with windings generates a back emf voltage "e" at the terminals of the phase. This voltage opposes the input voltage and is given by Faraday's law $e = d\lambda/dt$ [7].

The circuit can be expressed by the following equation:

$$V_0(t) = iR + \frac{d\lambda}{dt} \tag{4.8}$$

The mechanical side can be modeled, as shown in Figure 4.28. Here, T_e is the torque generated by the electrical side of the motor. T_f models static friction, such as Coulomb friction, while T_b can account for any energy dissipation due to friction in the bearings of the motor. The torque generated by the electrical side works against the torques due to the static friction, viscous friction, and the load. The difference of these torques is used to accelerate the rotor to a mechanical rotational speed of ω_m. Using Newton's second law, we can write

$$T_e - T_L - T_f - T_b = J\dot{\omega}_m$$

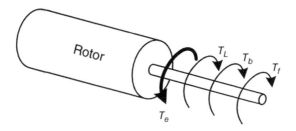

Figure 4.28   Mechanical model of a motor

where J is the rotor inertia. The viscous friction torque is often modeled as $T_f = b\,\omega_m$, where b is the viscous friction coefficient with Nm s/rad units. Substituting and rearranging give

$$T_e - T_L - T_f = J\dot{\omega}_m + b\,\omega_m \tag{4.9}$$

Also,

$$\omega_m = \frac{d\theta_m}{dt}$$

where θ_m is the rotor angular position (rad). Taking Laplace transform of both sides of Equation (4.9) with zero initial conditions yields

$$\omega_m(s) = \frac{1}{Js + b}\,(T_e - T_L - T_f) \tag{4.10}$$

The mechanical model in Equation (4.10) was implemented using Simulink® as shown in Figure 4.29a. This is a top-level block. When this block is opened, its details are revealed as shown in Figure 4.29b. The block takes inputs Te and TL as the torque generated by the electrical side and the load torque, respectively. It computes the angular speed wm and position theta. The signum function, Sign, was used to make sure that the static friction torque always opposes the input torque T_e. The J, B, and Tf values need to be specified for the simulated motor by entering them into MATLAB®.

4.5.1   AC Servo Motor Model

An AC servo motor receives phase voltages V_a, V_b, and V_c, which are then converted into torque by the stator and rotor field interactions. Starting from Equation (4.8), we can write the following equations for all phases of the motor:

$$V_a = i_a R + \frac{d\lambda_a}{dt}$$
$$V_b = i_b R + \frac{d\lambda_b}{dt} \tag{4.11}$$
$$V_c = i_c R + \frac{d\lambda_c}{dt}$$

Here, the flux linkage λ is defined as $\lambda = Li$, where L is the inductance and i is the current. The challenge is that the inductance is dependent on the rotor position and is not constant.

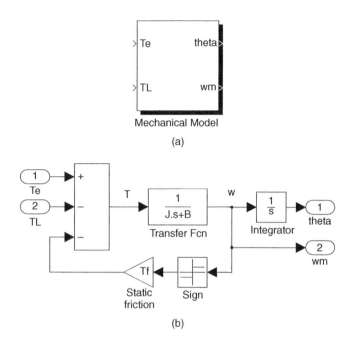

Figure 4.29 Mechanical model block in Simulink®. (a) Top-level Simulink® block for the mechanical model of a motor. (b) Second-level: Details of mechanical motor model block

This makes the analysis very complicated. A common approach is to transform the equations into a new coordinate frame known as the "*dq*-frame" where the inductances become constant (Figure 4.31). This transformation converts the three-phase system into an equivalent two-phase system. The *dq*-frame is attached to the rotor and rotates with it. The motor dynamic model turns into constant coefficient differential equations in the *dq*-frame.

Park transform is used to transform the three-phase stator quantities (such as phase voltages, currents, flux linkages) into the rotating *dq*-frame

$$\begin{Bmatrix} S_d \\ S_q \\ S_0 \end{Bmatrix} = \frac{2}{3} \begin{bmatrix} \cos\theta & \cos(\theta - 120) & \cos(\theta + 120) \\ -\sin\theta & -\sin(\theta - 120) & -\sin(\theta + 120) \\ 0.5 & 0.5 & 0.5 \end{bmatrix} \begin{Bmatrix} S_a \\ S_b \\ S_c \end{Bmatrix} \tag{4.12}$$

where S_a, S_b, and S_c are the stator quantities, and S_d and S_q are the corresponding *dq*-frame quantities. Angle θ is the electrical angle of the rotor. The S_0 term is included to yield a square matrix. In balanced three-phase systems, this component is ignored [7].

Inverse Park transform given in Equation (4.13) can be used to transform the *dq*-frame quantities back to three-phase quantities.

$$\begin{Bmatrix} S_a \\ S_b \\ S_c \end{Bmatrix} = \begin{bmatrix} \cos\theta & -\sin\theta & 1 \\ \cos(\theta - 120) & -\sin(\theta - 120) & 1 \\ \cos(\theta + 120) & -\sin(\theta + 120) & 1 \end{bmatrix} \begin{Bmatrix} S_d \\ S_q \\ S_0 \end{Bmatrix} \tag{4.13}$$

By applying the Park transform to the stator currents $[i_a \ i_b \ i_c]^T$ and the flux linkages $[\lambda_a \ \lambda_b \ \lambda_c]^T$, Equation (4.11) can be transformed into the dq-frame, where ω_e is the rotor electrical speed [20]

$$V_d = i_d R + \frac{d\lambda_d}{dt} - \omega_e \lambda_q \tag{4.14}$$

$$V_q = i_q R + \frac{d\lambda_q}{dt} + \omega_e \lambda_d \tag{4.15}$$

Figure 4.26b can then be redrawn as shown in Figure 4.30. The Park transform converts the stationary three-phase ABC coil system into two equivalent coils, L_d and L_q. These hypothetical coils are attached to the dq-frame, which rotates with the rotor (Figure 4.31). Therefore, the position dependent inductance L becomes constant L_d and L_q. These are called direct-axis (L_d) and quadrature-axis (L_q) inductances. Similarly, currents i_d and i_q are called the d-axis and q-axis currents, respectively.

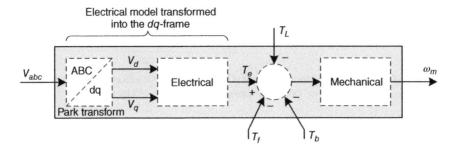

Figure 4.30   Expanded AC servo motor conceptual model

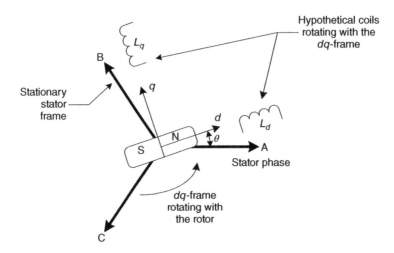

Figure 4.31 dq-frame and the hypothetical rotating direct-axis (L_d) and quadrature-axis (L_q) inductances

Along the d-axis, we have flux linkage due to the i_d current and the rotor permanent magnet (PM) as

$$\lambda_d = L_d i_d + \lambda_{PM} \tag{4.16}$$

Along the q-axis, we have flux linkage due to the i_q current as

$$\lambda_q = L_q i_q \tag{4.17}$$

Substituting Equations (4.16) and (4.17) into (4.14) gives

$$V_d = i_d R + \frac{d}{dt}(L_d i_d + \lambda_{PM}) - \omega_e L_q i_q$$

$$= i_d R + L_d \frac{di_d}{dt} - \omega_e L_q i_q$$

or

$$\frac{di_d}{dt} = \frac{1}{L_d}[V_d - i_d R + \omega_e L_q i_q] \tag{4.18}$$

Taking Laplace transform of both sides with zero initial conditions gives

$$i_d(s) = \frac{1}{L_d s}[V_d - i_d R + \omega_e L_q i_q] \tag{4.19}$$

Similarly, substituting Equations (4.16) and (4.17) into (4.15) yields

$$V_q = i_q R + \frac{d}{dt}(L_q i_q) + \omega_e L_d i_d + \omega_e \lambda_{PM}$$

or

$$\frac{di_q}{dt} = \frac{1}{L_q}[V_q - i_q R - \omega_e(L_d i_d + \lambda_{PM})] \tag{4.20}$$

Again, taking Laplace transform gives

$$i_q(s) = \frac{1}{L_q s}[V_q - i_q R - \omega_e(L_d i_d + \lambda_{PM})] \tag{4.21}$$

The torque generated by the motor is given by [7]

$$T_e = \frac{3}{2}\left(\frac{P}{2}\right)[\lambda_d i_q - \lambda_q i_d] \tag{4.22}$$

By substituting Equations (4.16) and (4.17) into (4.22) we obtain

$$T_e = \frac{3}{2}\left(\frac{P}{2}\right)[(L_d i_d + \lambda_{PM})i_q - (L_q i_q)i_d]$$

Rearranging the equation gives

$$T_e = \frac{3}{2}\left(\frac{P}{2}\right)[(L_d - L_q)i_d i_q + \lambda_{PM} i_q] \tag{4.23}$$

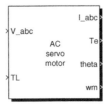

Figure 4.32   Top-level Simulink® model for a three-phase AC servo motor

The mechanical side of the motor is modeled by Equation (4.9) as explained earlier. It should be noted that the electrical speed ω_e and the mechanical speed ω_m are related to each other through the following equation where p is the number of poles

$$\omega_e = \left(\frac{p}{2}\right)\omega_m \tag{4.24}$$

To summarize, Equations (4.18), (4.20), and (4.23) model the electrical side of the motor while Equation (4.9), along with (4.22), models the mechanical side.

4.5.1.1   Simulation of an AC Servo Motor

The simulation model of the AC servo motor was built using Simulink®. This model will be used later in Chapter 6 to study control systems implemented in the drives.

The model is organized hierarchically. The top-level AC servo motor model is shown in Figure 4.32. If we open this top-level model, we see the details of the second-level, as shown in Figure 4.33. The `Electrical Model` block implements Equations (4.19) and (4.21) as shown in Figure 4.34a. The torque generated by the motor is given by Equation (4.23) which is computed as shown in Figure 4.34b.

The `abc->dq` and `dq->abc` blocks, as shown in Figure 4.33, implement the Park and inverse Park transforms given in Equations (4.12) and (4.13), respectively. The `abc->dq` block converts the ABC phase voltages (V_a, V_b, V_c) into dq-frame equivalents (V_d, V_q) before the motor model equations are computed. After the computation, the resulting dq-frame currents, i_d and i_q, are converted back into three-phase currents using the `dq->abc`. Details of these blocks are given in Figure 4.35. Each `f(u)` block in these figures is a user-defined function in Simulink®. They were used to define each row of the matrices in the Park transforms. For example, the first row of the matrix in Equation (4.12) was implemented by entering the following formula into the top `f(u)` block in Figure 4.35a:

$$u(1) * \cos(u(4)) + u(2) * \cos(u(4) - 2 * pi/3) + u(3) * \cos(u(4) + 2 * pi/3)$$

where $u(1)$, $u(2)$, and $u(3)$ are V_a, V_b, and V_c, respectively, and $u(4)$ is θ. Before running the simulation, values for the motor electrical parameters R, Ld, Lq, p, and PM (for λ_{PM}) need to be specified for the simulated motor by entering them into MATLAB®.

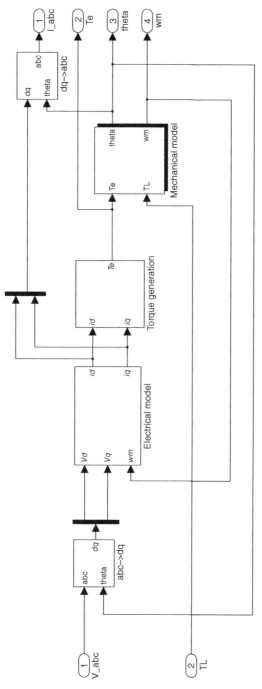

Figure 4.33 Second-level Simulink® model for a three-phase AC servo motor

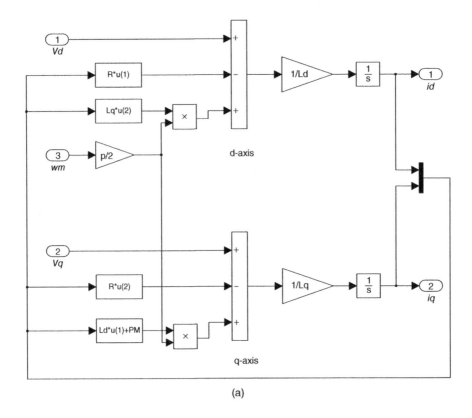

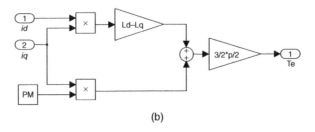

Figure 4.34 Electrical and torque generation models for AC servo motor. (a) Electrical model block details. Variables u(1) and u(2) correspond to i_d and i_q, respectively. (b) Torque generation block details

4.5.2 AC Induction Motor Model

An AC induction motor has a squirrel cage rotor and a three-phase stator. The rotating magnetic field of the stator induces currents in the conductors of the rotor which then generate the rotor magnetic field. The two fields interact to produce torque. Similar to the case of the AC servo motor, each phase of the AC induction motor can be modeled as shown in Figure 4.27 and Equation (4.11). However, using a vector format makes it easier to keep track of the resulting set of differential equations. Therefore, we will start with the following vector equation in the

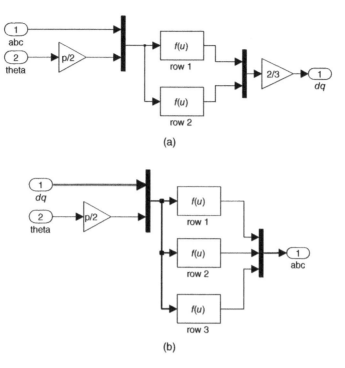

(a)

(b)

Figure 4.35 Park and inverse-Park transforms in the AC servo motor model. (a) abc-dq block details. (b) dq-abc block details

machine stationary frame (indicated by the superscript s)

$$\vec{V}_s^s = R_s \vec{i}_s^s + \frac{d\vec{\lambda}_s^s}{dt} \tag{4.25}$$

where $\vec{V}_s^s = [V_{as}\ V_{bs}\ V_{cs}]^T$, $\vec{\lambda}_s^s = [\lambda_{as}\ \lambda_{bs}\ \lambda_{cs}]^T$, and $\vec{i}_s^s = [i_{as}\ i_{bs}\ i_{cs}]^T$ are the instantaneous voltage, flux linkage, and current vectors expressed in the stationary stator frame, respectively. The subscript s indicates stator quantity such as the phase A stator voltage V_{as}. As before, the challenge is that inductance is dependent on the rotor position and is not constant. The common approach is to transform the equations written in the "ABC" stationary stator frame into a new "k" or "dq" coordinate frame. This transformation makes the inductances constant. There are several possibilities for the choice of the "k" frame. It can be selected as the stationary stator frame where the original three-phase system gets converted into an equivalent two-phase system. The rotor frame can also be selected where the "k" or dq-frame is attached to the rotor and spins with it. Another possible choice is the synchronous frame associated with the rotating magnetic field of the stator where the dq-frame rotates with the field. The choice of frame selection is related to the control strategies. If the reference system is rotating at ω_k speed, Equation (4.25) becomes [8]

$$\vec{V}_s = R_s \vec{i}_s + \frac{d\vec{\lambda}_s}{dt} + \vec{\omega}_k \otimes \vec{\lambda}_s$$

The additional term is called *speed voltage* and is due to the rotation of the reference frame. The cross-product $\otimes$ in this equation can be replaced by a rotation matrix $[R]$ and dot-product $\odot$ as follows [23]

$$\vec{V}_s = R_s \vec{i}_s + \frac{d\vec{\lambda}_s}{dt} + \omega_k [R] \odot \vec{\lambda}_s, \qquad \text{where} \quad [R] = \begin{bmatrix} 0 & -1 \\ 1 & 0 \end{bmatrix} \qquad (4.26)$$

Similarly, the electrical equation for the rotor is

$$\vec{V}_r = R_r \vec{i}_r + \frac{d\vec{\lambda}_r}{dt} + (\omega_k - \omega_{re})[R] \odot \vec{\lambda}_r \qquad (4.27)$$

where ω_{re} is the rotor electrical speed. Since the "k" frame is moving at ω_k speed, the rotor's relative speed with respect to this frame is $(\omega_k - \omega_{re})$. Each of the vector quantities in these transformed equations has a component along the d-axis and q-axis of the rotating frame. Therefore, Equations (4.26) and (4.27) can be rewritten as

$$\begin{Bmatrix} V_{ds} \\ V_{qs} \end{Bmatrix} = R_s \begin{Bmatrix} i_{ds} \\ i_{qs} \end{Bmatrix} + \frac{d}{dt} \begin{Bmatrix} \lambda_{ds} \\ \lambda_{qs} \end{Bmatrix} + \omega_k \begin{bmatrix} 0 & -1 \\ 1 & 0 \end{bmatrix} \begin{Bmatrix} \lambda_{ds} \\ \lambda_{qs} \end{Bmatrix}$$

$$\begin{Bmatrix} V_{dr} \\ V_{qr} \end{Bmatrix} = R_r \begin{Bmatrix} i_{dr} \\ i_{qr} \end{Bmatrix} + \frac{d}{dt} \begin{Bmatrix} \lambda_{dr} \\ \lambda_{qr} \end{Bmatrix} + (\omega_k - \omega_{re}) \begin{bmatrix} 0 & -1 \\ 1 & 0 \end{bmatrix} \begin{Bmatrix} \lambda_{dr} \\ \lambda_{qr} \end{Bmatrix} \qquad (4.28)$$

The flux linkages are defined as

$$\vec{\lambda}_s = L_s \vec{i}_s + L_m \vec{i}_r$$
$$\vec{\lambda}_r = L_m \vec{i}_s + L_r \vec{i}_r \qquad (4.29)$$

where L_s, L_r, and L_m are the stator, rotor, and magnetizing inductances, respectively. The stator inductance is made up of the magnetizing and self-inductance of the stator $(L_s = L_m + L_{sl})$. Similarly, the rotor inductance is made up of the magnetizing and self-inductance of the rotor $(L_r = L_m + L_{rl})$. These equations can be written in a matrix form as

$$\begin{Bmatrix} \vec{\lambda}_s \\ \vec{\lambda}_r \end{Bmatrix} = \begin{bmatrix} L_s & L_m \\ L_m & L_r \end{bmatrix} \begin{Bmatrix} \vec{i}_s \\ \vec{i}_r \end{Bmatrix}$$

Solving for the currents gives

$$\begin{Bmatrix} \vec{i}_s \\ \vec{i}_r \end{Bmatrix} = \begin{bmatrix} L_s & L_m \\ L_m & L_r \end{bmatrix}^{-1} \begin{Bmatrix} \vec{\lambda}_s \\ \vec{\lambda}_r \end{Bmatrix}$$

We can calculate inverse of the matrix and write the solution as follows:

$$\vec{i}_s = C_1 \vec{\lambda}_s - C_2 \vec{\lambda}_r$$
$$\vec{i}_r = -C_3 \vec{\lambda}_s + C_4 \vec{\lambda}_r \qquad (4.30)$$

where the coefficients are found as $C_1 = L_r/\Delta, C_2 = L_m/\Delta, C_3 = L_m/\Delta, C_4 = L_s/\Delta$ and, $\Delta = L_s L_r - L_m^2$.

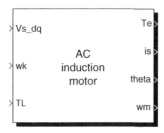

Figure 4.36 Top-level Simulink® model for a three-phase AC induction motor

The electrical torque produced by the motor can be formulated in various ways [23]. One definition is

$$
\begin{aligned}
T_e &= \frac{3}{2}\left(\frac{p}{2}\right)\left(\vec{i}_r \otimes \vec{\lambda}_r\right) \\
&= \frac{3}{2}\left(\frac{p}{2}\right)L_m\left(\vec{i}_r \otimes \vec{i}_s\right) \\
&= \frac{3}{2}\left(\frac{p}{2}\right)L_m\left([R]\vec{i}_r \odot \vec{i}_s\right)
\end{aligned}
\tag{4.31}
$$

To summarize, the equation set (4.28) and Equation (4.30) model the electrical side of the AC induction motor, while Equation (4.9), along with (4.31), models the mechanical side.

4.5.2.1 Simulation of an AC Induction Motor

The simulation model of the induction motor was built using Simulink®. This model will be used later in Chapter 6 to study control systems implemented in the drives.

The model is organized hierarchically. The top-level induction motor model is shown in Figure 4.36.

The model requires stator voltages V_{dq}^s in the dq-frame, electrical speed ω_k of the dq-frame, and the load torque as inputs. It then computes the motor electrical torque T_e, stator currents, i_s, rotor position θ, and rotor mechanical speed ω_m. The V_{dq}^s and i_s signals are vector quantities corresponding to those in Equation (4.28).

Opening this top-level model reveals its internal organization at the second-level as shown in Figure 4.37. If we take Laplace transform of the first Equation in (4.28) and re-arrange it, we can put it in the following form

$$
\left\{\begin{matrix}\lambda_{ds}\\\lambda_{qs}\end{matrix}\right\} = \frac{1}{s}\left(\left\{\begin{matrix}V_{ds}\\V_{qs}\end{matrix}\right\} - R_s\left\{\begin{matrix}i_{ds}\\i_{qs}\end{matrix}\right\} - \omega_k\begin{bmatrix}0 & -1\\1 & 0\end{bmatrix}\left\{\begin{matrix}\lambda_{ds}\\\lambda_{qs}\end{matrix}\right\}\right)
\tag{4.32}
$$

The upper left corner of the diagram in Figure 4.37 computes this equation. Similarly, the bottom equation in (4.28) can be re-arranged as

$$
\left\{\begin{matrix}\lambda_{dr}\\\lambda_{qr}\end{matrix}\right\} = \frac{1}{s}\left(\left\{\begin{matrix}V_{dr}\\V_{qr}\end{matrix}\right\} - R_r\left\{\begin{matrix}i_{dr}\\i_{qr}\end{matrix}\right\} - (\omega_k - \omega_{re})\begin{bmatrix}0 & -1\\1 & 0\end{bmatrix}\left\{\begin{matrix}\lambda_{dr}\\\lambda_{qr}\end{matrix}\right\}\right)
\tag{4.33}
$$

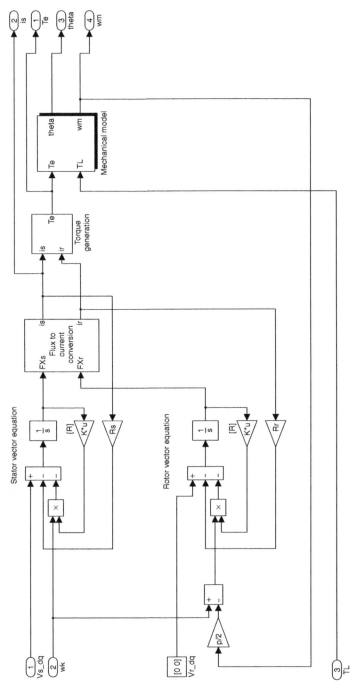

Figure 4.37   Second-level model: inside the AC induction motor

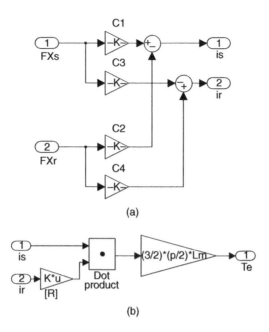

(a)

(b)

Figure 4.38 Computation of currents and torque generation in an AC induction motor. (a) Details of the "Flux to current conversion" block. (b) Details of the "Torque Generation" block

The lower left corner of the diagram in Figure 4.37 computes this equation. It should be noted that the Vs_dq, FXs, and FXr signals are vector quantities corresponding to the $\{V_{ds}\ V_{qs}\}^{\mathrm{T}}$, $\{\lambda_{ds}\ \lambda_{qs}\}^{\mathrm{T}}$ and $\{\lambda_{dr}\ \lambda_{qr}\}^{\mathrm{T}}$ vectors in these equations, respectively. Simulink® stacks the variables in each of these vectors into a 2×1 array and carries them through the computation as if they were signals with only one variable. The $\{V_{dr}\ V_{qr}\}^{\mathrm{T}}$ vector is set to $[0\,0]^{\mathrm{T}}$ since the model is simulating a squirrel cage induction motor where no external voltage is applied to the rotor windings. The rotor contains bars that are shorted by end caps. The matrix multiplication, $[R] \odot \vec{\lambda}_s$, is handled using gain matrix blocks in Simulink® shown as the triangular blocks with K*u in them.

Figure 4.38a shows the details of the Flux to current conversion block. It computes Equation (4.30). The torque generated by the motor is given by Equation (4.31) and is computed as shown in Figure 4.38b. Coefficients $C1, C2, C3, C4$ are computed using element-wise multiplications by the gain blocks with -K- in them.

The Mechanical Model block was explained earlier. It implements Equation (4.10) with details as shown in Figure 4.29.

Problems

1. What is a pole? What is a stator phase? How are the N and S Poles of a phase created?
2. In a motor with eight poles, how many mechanical degrees is one electrical cycle?
3. What are torque angle and commutation?

4. Figure 4.7 shows clockwise (CW) rotation of the rotor of a three-phase motor using six-step commutation algorithm. Make the same table as Table 4.1 but for counterclockwise (CCW) commutation of the same motor.

5. What are the primary components of an AC servo motor? What differentiates it from a brushless DC motor (BLDC)?

6. What is the synchronous speed of a three-phase motor with 6 poles operated from 60 Hz power source?

7. Define pole pitch, full-pitch winding, and fractional-pitch winding.

8. Given the fractional-pitch winding in Figure 4.39, make a plot for the flux linkage λ. Use the e_a back emf plot to infer the shape of the flux linkage plot. It should be similar to the λ plot in Figure 4.13.

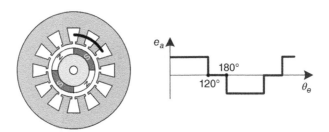

Figure 4.39   Fractional pitch coil and the resulting back emf waveform for Problem 8

9. What is torque ripple? What causes it?

10. What is the primary difference of a three-phase AC induction motor from a three-phase AC servo motor?

11. What is slip in an AC induction motor?

12. What is the primary difference of an inverter-duty or vector-duty motor from regular AC induction motors?

13. How does a VFD vary the speed of an AC motor while keeping the torque constant?

14. Why does the torque reduce in constant horsepower range of an AC induction motor with a drive?

References

[1] Step Motor and Servo Motor Systems and Control. Compumotor, Parker, Inc., 1997.

[2] App Note 3414: Sinusoidal Commutation of Brushless Motors. Technical Report, Galil, Inc., 2002.

[3] Application Note: Setup for PMAC Commutation of Brushless PMDC (AC-Servo) Motors and Moving Coil Brushless Linear Motors. Technical Report, Delta Tau Data Systems Inc., 2006.

[4] NEMA Standards Publication Condensed MG 1-2007, Information Guide for General Purpose Industrial AC Small and Medium Squirrel-Cage Induction Motor Standards, 2007.

[5] Brushless Motor Commutation. www.motion-designs.com, May 2008.

[6] ANSI/NEMA MG 1-2011, American National Standards, Motors and Generators. Technical Report, National Electrical Manufacturers Association, 2011.

[7] Stephen D. Umans A.E. Fitzgerald, Charles Kingsley Jr. *Electric Machinery*. McGraw-Hill Publishing Company, 1990.

[8] B. K. Bose. *Power Electronics and AC Drives*. Prentice-Hall, 1986.

[9] Sabri Cetinkunt. *Mechatronics*. John Wiley & Sons, Inc., 2006.

[10] Bill Drury. *The Control Techniques Drives and Controls Handbook*. Number 35 in IEE Power Series. The Institute of Electrical Engineers, 2001.

[11] Mohamed A. El-Sharkawi. *Electric Energy: An Introduction*. CRC Press, Taylor & Francis Group, Third edition, 2013.

[12] Emerson Industrial Automation. (2014)   Servo Motors Product Data, Unimotor HD.   http://www.emersonindustrial.com/en-EN/documentcenter/ControlTechniques/Brochures/CTA/BRO_SRVMTR_1107.pdf (accessed April 2014).

[13] Augie Hand. *Electric Motor Maintenance and Troubleshooting*. McGraw-Hill Companies, Second edition, 2011.

[14] Duane Hanselman. *Brushless Motors Magnetic Design, Performance, and Control*. E-Man Press LLC, 2012.

[15] Leslie Sheets James Humphries. *Industrial Electronics*. Breton Publishers, 1986.

[16] R. Krishnan. *Permanent Magnet Synchronous and Brushless DC Motor Drives*. CRC Press, Taylor & Francis Group, 2010.

[17] Lester B. Manz. The Motor Designer's Point of View of an Adjustable Speed Drive Specificaion. *IEEE Industry Applications Magazine*, pages 16–21, January/February 1995.

[18] Marathon Motors. (2014) Three Phase Inverter (Vector) Duty Black Max® 1000:1 Constant Torque, TENV Motor. http://www.marathonelectric.com/motors/index.jsp (accessed 5 November 2014).

[19] James Robert Mevey. Sensorless Field Oriented Control of Brushless Permanent Magnet Synchronous Motors. Master's thesis, Kansas State University, 2006.

[20] Salih Baris Ozturk. Modeling, Simulation and Analysis of Low-Cost Direct Torque Control of PMSM Using Hall-Effect Sensors. Master's thesis, Texas A&M University, 2005.

[21] Frank D. Petruzella. *Electric Motors and Control Systems*. McGraw-Hill Higher Education, 2010.

[22] H. Polinder, M. J. Hoeijmakers, and M. Scuotto. Eddy-current losses in the Solid Back-Iron of PM Machines for different Concentrated Fractional Pitch Windings. *IEEE International Electric Machines and Drives Conference*, 1:652–657, 2007.

[23] M. Riaz. (2012) Simulation of Electrical Machine and Drive Systems Using MATLAB and SIMULINK. University of Minnesota. http://www.ece.umn.edu/users/riaz/ (accessed 7 April 2012).

[24] Suad Ibrahim Shahl. (2009) Introduction to AC Machines: Electrical Machines II. http://www.uotechnology.edu.iq/dep-eee/lectures/3rd/Electrical/Machines%202/I_Introduction.pdf (accessed 13 April 2012).

[25] Siemens Industry, Inc. (2008) Basics of AC Motors. http://cmsapps.sea.siemens.com/step/pdfs/ac_motors.pdf.

[26] F. J. Teago. The Nature of the Magnetic Field Produced by the Stator of a Three-phase Induction Motor, with Special Reference to Pole-changing Motors. *Journal of the Institution of Electrical Engineers*, 61(323): 1087–1096, 1923.

[27] Yasuhito Ueda, Hiroshi Takahashi, Toshikatsu Akiba, and Mitsunobu Yoshida. Fundamental Design of a Consequent-Pole Transverse-Flux Motor for Direct-Drive Systems. *IEEE Transactions on Magnetics*, 49(7), 2013.

[28] Padmaraja Yedamale. AN885: Brushless DC (BLDC) Motor Fundamentals. Technical Report, Microchip Technology Inc. Technology Inc., 2003.

5

Sensors and Control Devices

Motion control systems employ various sensors and control components along with the motion controller. Sensors are devices to detect or measure a physical quantity such as position, temperature, or pressure. A sensor converts its input into a functionally related output. The input is the physical quantity and the output is usually an electrical signal. There are many types of sensors used in motion control applications. Central to the operation of any motion control system is the need to measure the position and speed of the load in motion. The most common sensor for this purpose is an optical encoder.

Control devices are used in building user interfaces and to govern the power delivery to electrical loads such as motors. Devices such as push buttons, selector switches, and pilot lights are commonly used in building the user interface on the control panel of an automated system. Devices such as contactors and overload relays are used in control circuits to operate high voltage motors.

This chapter starts with the presentation of various types of optical encoders for position measurement. Two techniques for velocity estimation from encoder data are discussed. Then, limit switches, proximity sensors, photoelectric sensors, and ultrasonic sensors used for the detection of objects are explained. Sinking or sourcing designations for sensor compatibility to I/O cards are explored. Next, control devices including push buttons, selector switches, and indicator lights are presented. The chapter concludes with an overview of motor starters, contactors, OLs, soft-starters, and a three-wire motor control circuit.

5.1 Optical Encoders

An encoder is a sensor that converts rotational or linear displacement into digital signals. As shown in Figure 5.1, a typical optical encoder consists of: (1) light source (LED), (2) code disk, (3) stationary mask, (4) light detector (photodetector), and (5) signal conditioning circuit. The disk sits between the light source and the photodetector. As the disk rotates, the pattern on it blocks or lets the light pass through the disk. The stationary mask ensures proper alignment of the light with the photodetectors. As a result, the detector signal turns on or off.

There are two main types of encoders: (1) incremental and (2) absolute. They follow the same operating principle but use different patterns on the disk.

Industrial Motion Control: Motor Selection, Drives, Controller Tuning, Applications, First Edition. Hakan Gürocak.
© 2016 John Wiley & Sons, Ltd. Published 2016 by John Wiley & Sons, Ltd.

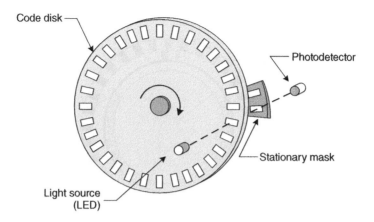

Code disk

Photodetector

Stationary mask

Light source
(LED)

Figure 5.1 Main components of an optical encoder (signal conditioning circuit not shown)

5.1.1 Incremental Encoder

A rotary incremental encoder produces a series of pulses as the disk rotates. Figure 5.2a shows a typical disk with one track. The black-and-white pattern is evenly distributed along the circumference of the disk. These disks can be mounted on various hubs (Figure 5.2b). The disk is then inserted in a detector unit to complete the core of the encoder assembly as shown in Figure 5.2c.

Figure 5.3 shows the digital pulses generated by a single detector as the disk pattern passes by it. A circuit in the motion controller can count the number of pulses as the shaft rotates. If the number of counts per revolution is known, then the motion controller can convert the accumulated pulse count into angular distance traveled by the shaft. However, we cannot tell whether the shaft is rotating clockwise or counterclockwise. In either direction, the controller would get pulses that look the same.

This problem can be solved if a second detector is added as shown in Figure 5.3. The detectors can be placed 90° E apart. One electrical cycle of the digital pulses consists of one high and one low pulse. One electrical cycle is equal to 360° E. Hence, when the two detectors are placed 90° E apart and side by side, each will point to a quarter of the electrical cycle. This arrangement creates the so-called *quadrature* signal pattern out of two channels CH A and CH B. When the disk moves, the alternating pattern seen by the detectors generates two square waveforms that are shifted in phase by 90° E.

One approach to determine the direction of rotation is to monitor which channel detects the first edge transition or change in its signal. Figure 5.4 shows an arbitrary starting position where both detectors happen to be pointing at the black area; hence, both signals are low. If the disk moves right, the CH A signal will change to high first. If the disk moves left, the CH B signal will transition to high first. Hence, by checking which channel senses the first signal change, we can tell if the disk is moving right or left.

If each black area is considered to be a radial line on the disk, going along the circumference from the leading edge of one black line to the leading edge of the next black line is a period of the waveform observed in one of the channels. As shown in Figure 5.5, there are four edge transitions in one period.

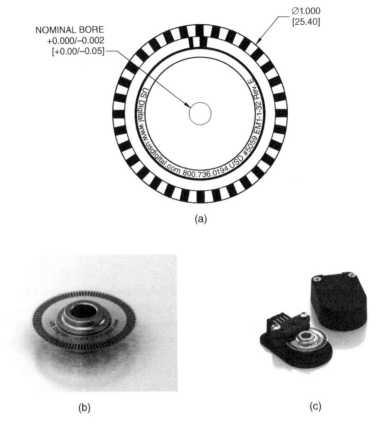

(a)

(b)   (c)

Figure 5.2 Incremental encoder (Reproduced by permission of US Digital Corp. [28]). (a) Disc with a single track black-and-white pattern. (b) Disc mounted on a hub. (c) Disc and detector assembly

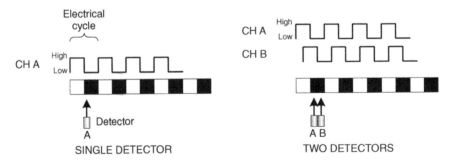

Figure 5.3 Encoder pulses from (a) single track with single detector and (b) single track with two detectors

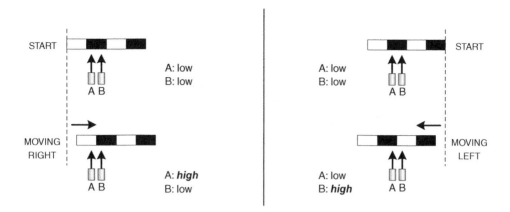

Figure 5.4   Determination of motion direction with two detectors

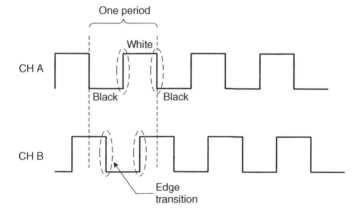

Figure 5.5   Quadrature signals

Therefore, if the motion controller circuit can count each edge transition, then the number of lines on an encoder disk can effectively be multiplied by 4. For example, if an encoder has 2000 lines, then 8000 counts will correspond to one revolution. In other words, this encoder will have a resolution of 360° M/8000 = 0.045°/ct. *Lines per revolution* (LPR) or *pulses per revolution* (PPR) are terms used in describing how many lines an incremental encoder disk has.

In Figure 5.2a, the main track with the black-and-white pattern is along the circumference of the disk. But on this disk, there is also a second track inside the main track. This track is all white except for a single mark at the top. This mark is called the *index pulse* which is often CH C on a typical encoder. This pulse is generated once per revolution. It is often used by motion controllers during home position search for an axis.

Incremental encoders are also available as linear encoders (linear scales) as shown in Figure 5.6. The operating principle is the same as the rotary encoders. However, the code disk

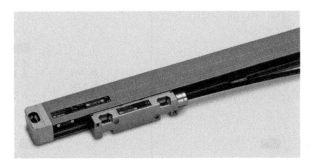

Figure 5.6   Linear encoder (linear scale) (Reproduced by permission of Heidenhain Corp. [10]))

is replaced by a linear strip with the code printed on it. They are used to measure displacement along a linear axis.

5.1.2   SinCos Encoders

SinCos (sine-cosine) encoders are incremental encoders with analog outputs. Unlike the discrete on/off outputs of the incremental encoders, the SinCos encoders provide two sinusoidal outputs shifted by 90° in phase. Hence, they are also known as *sinusoidal* encoders. The signals are generated using sinusoidal detectors. Industry standard for SinCos encoders is 1V peak-to-peak sine/cosine voltages [1, 4]. Due to the low voltage analog output signals, these encoders are sensitive to noise. As a result, complimentary channels are used for each signal. In addition, to avoid having to use negative power supply, typically 2.5 V DC offset is added to the signals as shown in Figure 5.7.

Each of the two analog channels provides a number of sine wave cycles per revolution as shown in Figure 5.7. For example, a SinCos encoder may provide ($1024 = 2^{10}$) sine wave cycles out of each channel in one revolution of its shaft. This corresponds to the lines per revolution in the incremental encoders.

A unique feature of the SinCos encoders is the ability to interpolate their output signals. This requires a special circuitry with analog-to-digital (A/D) converters in the motion controller. At a specific point in a single sine wave cycle (360° electrical), analog signals V_A and V_B are read from the channels. If channel A is the cosine channel and channel B is the sine channel, then arc tangent of the ratio of these two voltages will give the interpolated incremental *electrical* angle traveled within that sine wave cycle [8]. Hence,

$$\theta_{\text{interp}} = \tan^{-1}\left(\frac{V_B}{V_A}\right)$$

or, in counts

$$CTS_{\text{interp}} = \frac{2^n}{360} \tan^{-1}\left(\frac{V_B}{V_A}\right)$$

where n is the resolution of the A/D converter used in digitizing the analog channel signals. The arc tangent function $\tan^{-1}$ must be handled carefully by checking the sign of the sine and cosine signals to identify the correct quadrant for the angle.

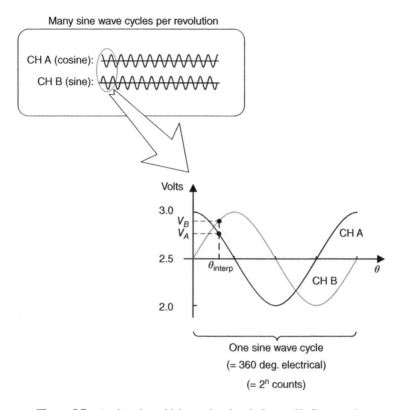

Figure 5.7   Analog sinusoidal encoder signals from a SinCos encoder

The circuitry also counts the number of sine wave cycles, n_{cycles}. Since each sine wave cycle can be digitized into 2^n bits by the A/D converter, the shaft position can be determined *in counts* as

$$CTS_{pos} = n_{cycles} \cdot 2^n + CTS_{interp}$$

■ **EXAMPLE 5.1.1**

A SinCos encoder provides 1024 ($= 2^{10}$) sine waves per revolution.

(a) If a 12-bit A/D is used by the motion controller circuitry, what will be the position resolution?
(b) If the position counter reads 4 193 102 *cts*, what is the mechanical angular position of the encoder shaft?

Solution
(a) Since the interpolation can resolve the angular position into 12-bits in each sine wave cycle, the overall position resolution will be

$$1024 \times 2^{12} = 4\,194\,304$$
$$= 2^{22}$$

With the 22-bit resolution the smallest detected position change will be

$$\frac{360}{2^{22}} = 8.583 \times 10^{-5^\circ} \text{ M}$$

(b) This encoder has 22-bit resolution, which is equal to 4 194 304 cts/rev. Then, the current shaft position in degrees can be found as

$$\theta = \frac{4\,193\,102}{4\,194\,304} \cdot 360$$
$$= 359.8968^\circ \text{ M}$$

5.1.2.1 Complimentary Channels

Electrical noise, such as due to the transients generated by the switching power electronics in the drive, can interfere with the encoder signals leading to wrong position counts. Shielded, twisted-pair cables need to be used and good wiring practices must be followed. An effective way to reduce the impact of noise is to use complimentary signals and *balanced differential line drivers* (EIA standard RS422). In this approach, each encoder channel has its complimentary channel, which is inverted as shown in Figure 5.8. Both quadrature and SinCos encoders come with complementary (differential) channels. They are often labeled as CH A+, CH A−, CH B+, CH B−, and CH C+, CH C− (or Index+, Index−). The inverted signals allow bigger voltage swings (such as 0–10 V instead of the usual 0–5 V outputs) and, therefore, are more immune to noise.

Line drivers are differential amplifiers. They increase the output current of each channel to "drive" the signal along the cable between the encoder and the controller. As shown in Figure 5.9, the differential line driver gets the signal from the encoder and inverts it to create the complimentary signal. These signals are then sent to the controller through the shielded, twisted-pair cable. Since the two wires are in the same environment, any electrical noise will affect both of them in a similar way. On the controller side, a differential receiver will invert the CH A− signal again and combine it with the CH A+ signal. The common-mode noise induced on the signals is rejected by the differential amplifier and the original CH A+ signal of the encoder is extracted. There are line drivers on each of the encoder channels.

5.1.3 *Absolute Encoder*

An incremental encoder can measure a relative distance by counting the pulses. If the counting is referenced to a specific point, such as a home position or the index pulse on

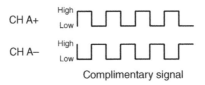

Figure 5.8 Complimentary (differential) encoder channels to increase noise immunity

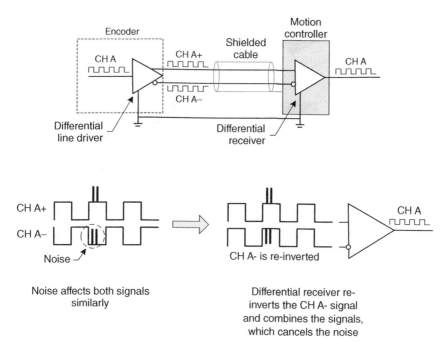

Figure 5.9   Noise immunity with differential line drivers

the encoder, then position can also be measured relative to that point. However, if there is a loss of power, the counts will be lost. Similarly, when the machine is first powered up, it is not possible to know the position of an axis with an incremental encoder before homing is completed.

An absolute encoder uses a disk that generates a unique digital code for each position of the disk. Therefore, the absolute position of the shaft can be determined at any time, even during power up. If an axis uses an absolute encoder, there is no need to move it to a known reference point (home position) in the power up sequence.

The resolution of an absolute encoder is determined by the number of bits in its output. Figure 5.10 shows 3-bit encoder disks. In this case, the encoder has three detectors, one on each track. The smallest position change that can be detected is $360°/2^3 = 45°$. Absolute encoders with 12-bits are common, which can detect $360°/2^{12} = 0.088°$.

The absolute encoder disk can have a *binary code* as shown in Figure 5.10a. Each sector (45° slice in this case) corresponds to the three-digit code in Table 5.1. However, due to small misalignment between the detectors during the manufacturing of the encoder, the bits of the binary code may not transition all at the same time as the disk rotates from one sector to the next. As a result, the encoder may appear to be jumping from the current track to another track which is not adjacent to the current one. To remedy this problem, *Gray code* is commonly used. It is much less likely to produce a wrong reading since only 1-bit at a time changes as the disk rotates from one sector to the next. Even if there were manufacturing imperfections, due to the single-bit change from sector to sector, the error in the position reading would be only as much as one sector (e.g., 0.088° in a 12-bit encoder).

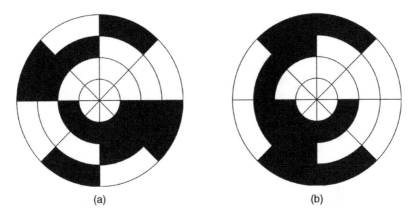

(a)   (b)

Figure 5.10   Absolute encoder disks with 3-bit resolution. (a) 3-bit binary code disc [2]. (b) 3-bit gray code disc [12]

Table 5.1   Absolute encoder disk codes (3-bits)[a]

Sector	Binary code			Gray code		
	D1	D2	D3	D1	D2	D3
0	0	0	0	0	0	0
1	0	0	1	0	0	1
2	0	1	0	0	1	1
3	0	1	1	0	1	0
4	1	0	0	1	1	0
5	1	0	1	1	1	1
6	1	1	0	1	0	1
7	1	1	1	1	0	0

[a]D1, D2, D3 are detectors on each track. Sector zero is horizontal from 0° to 45°. Sector number increases counter clockwise.

A *single-turn absolute encoder* can provide a unique code for each angular position within one revolution of the shaft. However, if the shaft rotates more than one revolution, then the codes will be repeated. A *multiturn absolute encoder* can count the number of revolutions and store them. It contains a gear train with high-precision gears. Each gear has its own code disk to count the turns. Multiturn encoders with 4-code disks are available. The code disks on the turn counters typically use the 4-bit Gray code (= 16 steps). Every time the main disk (first disk) completes a revolution, the second disk advances by one count. When the second disk finishes a complete revolution, the third disk advances by one count. Similarly, when the third disk finishes a complete revolution, the fourth disk advances by one count.

Figure 5.11   Gray code disk (9-bits) [16]

■ **EXAMPLE 5.1.2**

A multiturn absolute encoder has a main disk with a 9-bit Gray code as shown in Figure 5.11. It also has three 4-bit disks for turn counts. What is the resolution of this encoder?

Solution
Each 4-bit turn counter will have $2^4 = 16$ counts. Then, with the 9-bit main disk, this 4-disk multi-turn encoder will have 21-bit resolution

$$2^9 \cdot 16 \cdot 16 \cdot 16 = 2\,097\,152$$
$$= 2^{21}$$

5.1.4   Serial Encoder Communications

Incremental encoders output a series of pulses. On the other hand, absolute encoders can make all bits of the code for the absolute position available at once through a *parallel interface*. The disadvantage is that a cable with as many wires as the number of bits in the code disk is required. The rapid availability of the data in real time is, however, an advantage.

Several serial communication protocols have been developed to output absolute encoder data. Some of these protocols are proprietary, while the others are open protocols. Performance trade-offs include wire count, specific hardware requirements, and update rates.

5.1.4.1   SSI® (Synchronous Serial Interface)

Originally developed by Max Stegmann GmbH in 1984, SSI® is a popular, fundamental open serial communication protocol. As shown in Figure 5.12, there are two signal lines between the encoder and the controller: (1) clock and (2) data. These lines carry differential signals and use twisted-pair wires for better noise immunity. In addition, two wires are needed for the power and ground. Hence, the interface uses a total of six wires.

Initially, both the clock and the data lines are high. When the controller (master) needs position data from the encoder (slave), it starts to send a stream of clock pulses. At the first

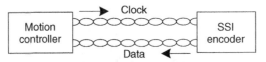

Two channels with twisted-pair wires

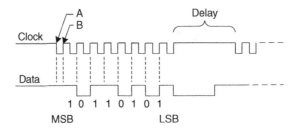

Figure 5.12   SSI interface

falling edge of the clock signal (point A), the encoder latches its current position data into its shift register. At the next rising edge of the clock signal (point B), the encoder starts sending its data. One bit is sent in each clock cycle starting with the most significant bit (MSB) first and finishing with the least significant bit (LSB). The SSI® interface of the controller must be programmed for the specific word length of the encoder. For example, for the 9-bit encoder, as shown in Figure 5.11, the controller would need to be programmed for 9-bit word length. Since it takes one extra clock cycle at the beginning to get the transmission started, the controller interface will add one more clock cycle to the transmitted clock pulses (10 clock pulses in this case). At the end of transmitting all data bits, the clock signal remains high for a minimum required time delay so that the encoder can recognize the end of transmission. After this delay, the next data transmission can begin. Typical data transmission rates of up to 1.5 MHz are possible but the cable length must get shorter as the frequency increases [1, 4].

5.1.4.2   EnDat® (Encoder Data)

This is a proprietary serial communication protocol by Heidenhain Corp. of Germany [9]. The EnDat® 2.2 is a bi-directional interface that is capable of transmitting position data from incremental and absolute encoders. It can also read information, such as encoder model number, serial number, direction of rotation, pulses per revolution, diagnostics, and alarm codes, stored in the encoder or update the stored information. The additional information is transferred along with the position value through the data channel. Transmission mode commands define the format of the data. The formats include position only, position with additional information, or just the parameters.

Just like the SSI® protocol, only four wires are used for the clock and data signals. Clock frequencies up to 8 MHz with up to 100 m of cable length are possible. The incremental encoder data channel is separate from the clock/data channels for the absolute encoder. Models with or without the incremental data channel are available.

5.1.4.3   HIPERFACE®

This proprietary protocol is by SICK/Stegmann Corp. The HIPERFACE® can be used point-to-point (one master and one slave) just like the SSI® and EnDat®. But it can also be used with bus connections where one master can address several encoders [25].

The encoder has a special code disk. One track contains a bar-code pattern that measures absolute position within one revolution. Absolute position with up to 15-bit resolution is available. A second track is a SinCos encoder. The use of analog sine/cosine signals for the incremental position measurement offers the ability to interpolate the signals for a very high resolution (up to 4 million cts/rev) at the lowest possible bandwidth. Typically, the absolute position is transmitted by the encoder to the controller when the system is first powered up. The controller can use this information in motor commutation. Subsequently, the incremental position values are transmitted from the sine/cosine channels and interpolated as explained in Section 5.1.2 to obtain very high resolution.

The HIPERFACE® also offers smart sensor capabilities to communicate electronic name-plate parameters such as motor voltage, current, and to store user parameters such as operating and maintenance history. As shown in Figure 5.13, the HIPERFACE® uses only eight wires. For high noise immunity, shielded twisted-pair cables must be used. Unlike the SSI® and EnDat®, the HIPERFACE® uses asynchronous protocol to send and receive data. The data rate is 38.4 kbps but contains lots of information.

5.1.4.4   BiSS® (Bi-directional Synchronous Serial Interface)

This is an open protocol by iC-Haus in Germany [11]. Unlike the 1 V peak-to-peak analog signals of the SinCos encoders, BiSS® is an all-digital approach that provides a noise immune link between the encoder and the controller. Normally, interpolation of the SinCos encoder data happens in the controller once the signals are received. Instead, the BiSS® encoders use internal interpolation, hence do not get affected from capacitive attenuation. Position updates at 10 MHz are possible, which enable very smooth motion even in low speeds.

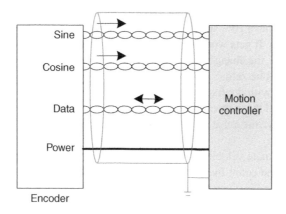

Figure 5.13   HIPERFACE interface

As with the SSI® encoders, BiSS® also uses four data wires, two for data and two for clock signals. There are also two wires for power. Like the HIPERFACE®, the BiSS® encoders can provide data such as resolution, device ID, acceleration, temperature, etc. These parameters can be transmitted bi-directionally between the master and the encoder. As in the case of the HIPERFACE® encoders, point-to-point or bus connection operations are possible. The cable length can be up to 150 m. The system is able to measure the transmission delay and compensate for it automatically. Therefore, the length of the cable does not influence the system dynamics [14].

5.1.5 Velocity Estimation

Drives used in earlier motion control systems employed tachometers for velocity measurement and to close the velocity loop. The position of the axis was measured using either an encoder or a resolver. In recent years, powerful computational capabilities of the drives and the shift to AC motors requiring commutation led to systems using solely position feedback. The velocity is estimated from the position data typically coming from an encoder.

An encoder quantizes the position with a limited resolution. For example, a 1000-line encoder with quadrature decoding would give 4000 cts/rev. Hence, the minimum position we can detect is $1/4000 = 0.00025$ revolutions. Controllers, sample position at a fixed rate called the *sampling period* t_s.

One way to estimate the velocity is by calculating the change in the position counts during the sampling period [27]. This approach is called the *fixed-time (FT) method* and is given by

$$V(k) = \frac{P(k) - P(k-1)}{t_s}$$

where $V(k)$ is the velocity estimate at the kth sampling instant. $P(k)$ and $P(k-1)$ are the position counts at the kth and $(k-1)$st samples, respectively. Let us assume that the motion controller runs at 2500 kHz sampling rate. This gives $t_s = 1/2500$ s. With this sampling period and the encoder mentioned earlier, the minimum velocity we can measure will be 37.5 rpm (details in Example 5.1.3). If the motor is running at 3750 rpm, this would mean 1% speed error due to the velocity estimation. On the other hand, if the motor is running at a low speed of 375 rpm, we would have 10% speed error coming from the quantization of the encoder and the velocity estimation. It gets worse when the motor is running slower than 37.5 rpm, since in some of the samples, the change in the position counts cannot be detected and the resulting velocity estimation will be zero.

Velocity estimation with the fixed-time method creates noise spikes in the velocity signal. The noise goes through the control system leading to current spikes to the motors. Low-pass filters are used to remove the noise but they end up causing phase-lag and forcing the controller gains to be lowered.

Another way to estimate velocity is by measuring the time elapsed between two consecutive encoder pulses (one count increase in the position). This method is often called the *1/T interpolation method* and is given by

$$V(k) = \frac{1}{T(k) - T(k-1)}$$

where $T(k)$ is the time at the kth encoder pulse. The denominator is the time elapsed between two consecutive encoder counts. This approach requires a highly accurate timer in the controller. Today's motion controllers have timers with 1 µs accuracy.

Since the time measurement has very high resolution, the velocity estimation with the $1/T$ interpolation method will be more accurate, especially at very low speeds [3]. The elapsed time between two encoder counts is proportional to the motor speed. If a very high resolution encoder is used with a motor running at high speeds, the time interval for one count will be very short. Depending on the accuracy of the timer, significant errors can be introduced into the velocity estimations at high speeds. In addition, the $1/T$ interpolation method adds phase lag, especially at low speeds, reducing the control loop performance [5].

■ EXAMPLE 5.1.3

Consider a motor with a 1000-line incremental encoder and quadrature decoder.

(a) Assume that the controller has 2500 kHz sampling rate and uses the fixed-time method for velocity estimation. What will be the speed estimation error if a one-count error was made in the position counts while the motor was running at the minimum detectable speed?

(b) Assume that the encoder is connected to a controller that uses the $1/T$ interpolation method with a 1 µs timer. If the motor is running at the same minimum speed as in the first case, what will be the speed estimation error if a 1-count error was made in the timer counts?

Solution

(a) At the given sampling rate, the sampling period will be $t_s = 1/2500 = 0.0004$ s. The velocity that leads to a change of one count in the position in this much time is the minimum velocity we can estimate using this method

$$V_{min} = \left(\frac{1\,\text{cts}}{0.0004\,\text{s}}\right)\left(\frac{1\,\text{rev}}{4000\,\text{cts}}\right)\left(\frac{60\,\text{s}}{1\,\text{min}}\right)$$
$$= 37.5\,\text{rpm}$$

Now, let us assume that the motor is running at this minimum speed of 37.5 rpm. Depending on how the sampling interval corresponds to the changes in the encoder counts, we may detect a change of one or two counts during the sampling period. If there is a one-count error in the position counts, then the controller will have two counts in this sampling interval

$$V = \left(\frac{2\,\text{cts}}{0.0004\,\text{s}}\right)\left(\frac{1\,\text{rev}}{4000\,\text{cts}}\right)\left(\frac{60\,\text{s}}{1\,\text{min}}\right)$$
$$= 75\,\text{rpm}$$

which is $\frac{(75-37.5)}{37.5} \times 100 = 100\%$ error in the velocity estimation !

(b) If the motor is running at 37.5 rpm, we will get 2500 cts/s ($= (37.5/60\,\text{rev/s}) \cdot 4000\,\text{cts/rev}$). The elapsed time for one count is 0.0004 s ($= 1/2500$). If the timer of the motion controller can count in µs intervals, the 0.0004 s will be 400 timer counts for one encoder position count. Let us now assume that there was an error in the timer counter by one count (401 cts instead of 400 cts). Then, the estimated velocity will be

$$V = \left(\frac{2\,\text{cts}}{0.000401\,\text{s}}\right)\left(\frac{1\,\text{rev}}{4000\,\text{cts}}\right)\left(\frac{60\,\text{s}}{1\,\text{min}}\right)$$
$$= 37.406\,\text{rpm}$$

which is only $\frac{(37.5-37.406)}{37.5} \times 100 = 0.25\%$ error in the velocity estimation.

5.2   Detection Sensors

In automated systems, there is a need to detect the presence or absence of an object. Detection sensors use a variety of technologies ranging from mechanical contact with the sensor to infrared light or ultrasonic signals to detect an object.

Regardless of the technology used, sensors that detect the presence or absence of an object act like an ordinary switch that is either open or closed. If a sensor is normally open (N.O.), it will close when it detects an object. Similarly, if it is normally closed (N.C.), it will open when it detects an object.

5.2.1   Limit Switches

One of the most common detection sensors found in automated systems is the limit switch. A mechanical limit switch is comprised of a switch and the operator (actuator). The most common switches have a contact block with one N.O. and one N.C. set of contacts. There are a variety of operators including lever-type, push roller-type, wobble stick, and fork-lever [15]. A mechanical switch detects the presence of an object when the object contacts the lever of the switch. Figure 5.14 shows a mechanical limit switch and typical symbols used in circuit diagrams to represent limit switches.

5.2.2   Proximity Sensors

Proximity sensors can detect the presence of an object without contacting the object. These sensors function like a switch (open/closed). They are available in two types: (1) inductive and (2) capacitive.

The inductive-type sensors can detect the presence of metallic (ferrous and nonferrous) targets. An oscillator in the sensor creates a high-frequency electromagnetic field at the tip of the sensor. When a metallic target enters the field, the oscillation amplitude decreases, which is detected by the signal conditioning circuit in the sensor and the "switch" is closed. The *operating range* of the sensor can be up to two inches. When a target gets in this range, the sensor detects it. Figure 5.15 shows a three-wire DC proximity inductive sensor. More information about the three-wire sensor wiring can be found in Section 5.2.6.

The capacitive proximity sensors look similar to those of the inductive type. They generate an electrostatic field and can detect both conductive and non-conductive targets. Two electrodes at the tip of the sensor form a capacitor. When a target gets near the tip, it changes

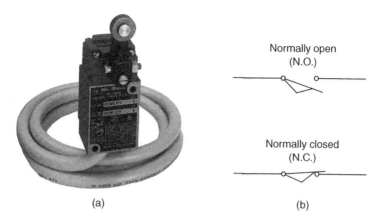

(a)   (b)

Figure 5.14 Mechanical limit switch and circuit symbols. (a) Mechanical limit switch with roller-lever operator (Courtesy of *Rockwell Automation, Inc.*) [17]. (b) Circuit symbols for N.O. and N.C. limit switches

(a)   (b)

Figure 5.15 Inductive proximity sensor and wiring. (a) Inductive proximity sensor (Courtesy of *Rockwell Automation, Inc.*) [18]. (b) Three-wire N.O. proximity sensor wired to a load (input card of a motion controller)

the capacitance and the oscillator output. The signal conditioning circuits in the sensor detect the change and close the "switch." The sensing range is typically shorter than the inductive sensors.

5.2.3 Photoelectric Sensors

These sensors employ light to detect the presence of objects and work like a switch. A photoelectric sensor contains a transmitter and a receiver. The transmitter is the light source which is often invisible infrared light. The receiver detects the light. If the sensor output is normally closed when the receiver detects light, it will open if the light is interrupted by an object.

Three methods of detection are available: (1) through-beam, (2) retro-reflective, and (3) diffuse sensing as shown in Figure 5.16. In the through-beam method, the transmitter and the receiver are aligned along a straight line facing each other. When an object interrupts the light beam between the receiver and the transmitter, the sensor changes its state.

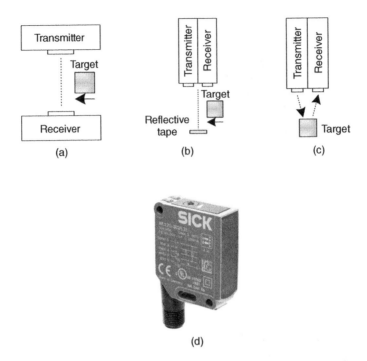

Figure 5.16 Photoelectric sensor detection methods and a diffuse mode industrial sensor. (a) Through-beam. (b) Retroreflective. (c) Diffuse sensing. (d) Diffuse mode infrared industrial photoelectric sensor [13]

In the retro-reflective method, the transmitter and the receiver are mounted next to each other on the same side of the beam. The light emitted from the transmitter gets reflected back from a reflective tape at the opposite side of the beam across the sensor. The reflected beam is then received by the detector. If the beam is interrupted, nothing gets reflected back to the detector, which then changes the output state of the sensor.

In the diffuse sensing approach, the transmitter and the receiver are enclosed in a single housing. The object reflects back the transmitter beam. The scattered light is then detected by the receiver. The scanning distance of these sensors is limited since they rely on the scattered light from the object.

5.2.4   Ultrasonic Sensors

An ultrasonic sensor such as the one shown in Figure 5.17 generates a high-frequency sound wave. If a target is in the range of the sensor, the sound is reflected back to the sensor. The sensors can produce two types of output: (1) discrete and (2) analog. Ultrasonic sensors can detect uneven surfaces, solids, clear objects, liquids, and course-granular materials. They are also popular in industrial web handling systems in measuring the web roll diameter for tension control.

Figure 5.17 Ultrasonic sensor (Courtesy of *Rockwell Automation, Inc.*) [19]

Sensors with analog output measure the elapsed time between a sound pulse from the sensor and the echo received back from an object. Since sound travels at a fixed speed, measuring the elapsed time allows us to calculate the distance to the object. Analog sensors have an operating range defined by a minimum and maximum distance they can measure. If an object is closer to the sensor than the minimum distance, it will be in the blind zone of the sensor and cannot be detected. Within the operating zone, the sensor produces an analog output signal directly proportional to the distance to the object. The output signal is either a current (4–20 mA) or voltage (0–10 V DC). There are many sensors available. Depending on the sensor, the range can be anywhere from 6–30 to 80–1000 cm [26].

Sensors with discrete output work like a switch. They are used to detect the presence or absence of an object. For example, if the sensor output is normally open, it will switch to a closed state when an object is detected in its range. Just like the photoelectric sensors (Figure 5.16), ultrasonic sensors can be used in the through-beam, retro-reflective, or diffuse modes. Programmable ultrasonic sensors are available that can work in either analog or proximity mode. Their outputs can also be programmed to be either N.O. or N.C.

5.2.5 The Concept of Sinking and Sourcing

There are many sensors available on the market. Sensors selected for a specific application must be electrically compatible with the machine I/O of the motion controller. Sinking or sourcing designations are used to describe the direction of current flow between the sensor and the I/O on the controller. The definitions are based on the assumption of the conventional DC current flow from positive to negative.

Consider a simple circuit made of a power supply, a switch, and a bulb, as shown in Figure 5.18. There are two ways of wiring this circuit with the switch before the bulb or after the bulb. Let us separate this circuit into its components to understand the concept of sourcing and sinking.

If the switch is connected to the positive side of the power supply as shown in Figure 5.18a, the switch becomes a *sourcing device*, since it *provides current* to the bulb. A sinking device must be connected to the switch so that a path for the current to return to the negative side of the power supply can be provided. Then, the bulb is a *sinking device* since it *receives current* from the switch.

If the switch is connected to the negative side of the power supply as shown in Figure 5.18b, then it becomes a sinking device since it receives current from the bulb. On the other hand,

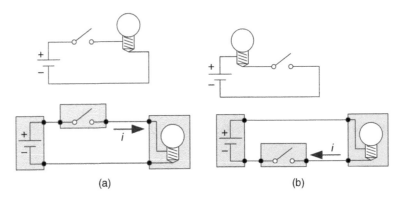

Figure 5.18 Two ways to wire a simple circuit with a switch. (a) Switch before the bulb. Switch is sourcing. (b) Switch after the bulb. Switch is sinking

the bulb now is a sourcing device since it provides current to the switch device. Again, the sourcing device has to be connected to a sinking device to complete the circuit.

5.2.5.1 Input Cards

We can now extend this simple circuit and the sinking/sourcing concept to applications with motion controllers. A more generalized term for the switch is *field device*, which can be any on/off sensor not just a switch. Similarly, we can generalize the bulb by replacing it with an *input card* in a motion controller. The input card has pins or screw terminals where field devices are connected. Recall that the bulb, as shown in Figure 5.18, turns on when the switch is closed. Similarly, a specific input (or pin) on the input card "turns on" when the switch is closed. In the case of the bulb, we can see it turning on. In the case of the input card, the motion controller software can "see" the input turning on.

Input cards can be sourcing or sinking type. In Figure 5.19a, the switch is a sourcing field device connected to a sinking input card. The switch provides (sources) current to the input card, which receives (sinks) it. In Figure 5.19b, the switch is a sinking field device connected to a sourcing input card. In this case, the input card provides (sources) current to the switch, which receives (sinks) it. In both cases, the input pin (IN) is turned on and recognized by motion controller software when the switch is closed.

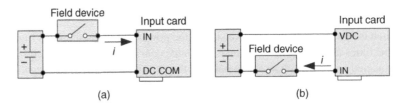

Figure 5.19 Sinking or sourcing input card. (a) Sinking input card, sourcing field device. (b) Sourcing input card, sinking field device

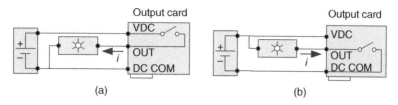

Figure 5.20   Sinking or sourcing output card. (a) Sourcing output card, sinking field device. (b) Sinking output card, sourcing field device

5.2.5.2   Output Cards

The sinking/sourcing concept continues to apply. An output card switches power on/off to an external device. The card does not supply power by itself. Instead, the power comes from an external power supply. Hence, a power supply must be connected to the output card.

In Figure 5.20a, the output card supplies (sources) current to the field device, which receives (sinks) the current. The output (switch) is turned on or off by software. A sourcing output card uses a PNP transistor. As shown in Figure 5.20b, the field device is sourcing and the output card is sinking. When the output is turned on by software, the card sinks the current to provide the return path to the negative side of the power supply. A sinking output card uses an NPN transistor.

5.2.6   Three-Wire Sensors

In Section 5.2.5, we looked at how field devices with two wires can be interfaced to sinking- or sourcing-type I/O cards. Although a switch was used in Figures 5.19 and 5.20 to explain the concept, any on/off sensor with two wires can be substituted for the switch in the figures.

Three-wire sensors use solid-state circuits which provide faster switching action with no contact bouncing unlike the mechanical switches. The internal "switch" circuits of these sensors use transistors. Therefore, power must be supplied to the sensor. The IEC60947-5-2 standard specifies wire colors for these sensors. Brown and blue wires are connected to the positive and negative side of the power supply, respectively. The black wire is the output signal of the sensor.

Figure 5.21a shows a three-wire sourcing-type sensor. A sourcing-type sensor uses a PNP transistor as the switching element. When the sensor is activated by an external event, the transistor is turned on, which then supplies current to the input card. The term "load" is often used in the literature from sensor manufacturers. It refers to the device the sensor is connected to. In this case, the load is the *sinking* input card of the motion controller.

Figure 5.21b is a three-wire sinking-type sensor. A sinking-type sensor uses an NPN transistor as the switching element. When the sensor is triggered, the transistor is turned on to receive current from a sourcing load. Therefore, the output card in this case must be of sourcing type. When the sensor turns on, current comes from the card into the sensor (transistor) and goes out from the DC COM terminal into the negative side of the power supply.

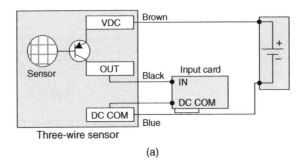

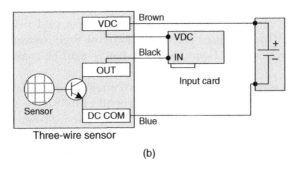

Figure 5.21 Three-wire sensor wiring. (a) Three-wire sourcing-type sensor (with PNP transistor). (b) Three-wire sinking-type sensor (with NPN transistor)

I/O INTERFACING RULES

---

1. A *sinking device* must be wired to a *sourcing card*
2. A *sourcing device* must be wired to a *sinking card*

5.3 Pilot Control Devices

A pilot control device, such as a push button, activates a *primary control device* such as a motor contactor. Pilot devices are used to construct user interface for a machine. A typical basic user interface consists of push buttons, selector switches, and indicator lights.

Pilot devices are available in three standard sizes based on the diameter of the mounting hole required on the panel: 16, 22.5, and 30.5 mm. The 30.5 mm size has been popular in the United States, whereas the 16 and 22.5 mm sizes originated from Europe (Figure 5.22). IEC (International Electrotechnical Commission) and NEMA (National Electrical Manufacturers Association) standards govern these devices.

(a)   (b)

Figure 5.22 IEC- and NEMA-style push buttons (Courtesy of *Rockwell Automation, Inc.*). (a) IEC style push button [20]. (b) NEMA style push button [23]

5.3.1 Push Buttons

Push buttons are used to manually open or close electrical contacts. They come in a variety of styles, colors, and features such as buttons with flush head, extended head, and with or without guards.

The button assembly consists of three parts: (1) operator, (2) legend plate, and (3) contact blocks. The *operator* is the part that is pressed, pulled, or twisted to activate the contacts. The *legend plate* is the label of the button (e.g., START). The *contact block* is where the electrical contacts are enclosed (Figure 5.23a). There are N.O. and N.C. contacts. In the case of the N.O. type, pressing the button closes the otherwise open contacts. The N.C. type works in the opposite way in that pressing the button opens the contacts that are otherwise closed. The electrical contacts in the contact block are spring loaded. They return to their normal state when the operator is released.

Operators are available for *momentary-contact* or *maintained-contact* operation. Momentary push button operators are spring loaded. As soon as it is released, the operator goes back to its normal state. For example, if a momentary push button operator is mounted on an N.O. contact block, then we will have a switch that is normally open. When the button (the operator) is pressed, the contacts will be closed. As soon as the button (operator) is released, the contacts will be opened. If the same operator is mounted on an N.C. contact, then we will have a switch that is normally closed. Pressing the button will open the contacts. When released, the contacts will close again. Figure 5.23b shows symbols used in wiring diagrams for these types of buttons.

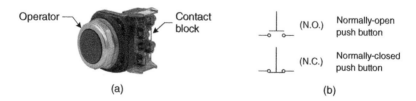

(a)   (b)

Figure 5.23 Industrial push button operator, contact block and symbols. (a) Operator and contact block Photo: (Courtesy of *Rockwell Automation, Inc.*) [23]. (b) Symbols for push button

Figure 5.24 Push/pull emergency stop button (Courtesy of *Rockwell Automation, Inc.*) [21]

The maintained-contact operators remain in their activated state. For example, an E-STOP button will remain pressed even after it is released. This maintains the state of the contacts until the button is physically reset. Typical two-position E-STOP buttons have either push/pull-release (Figure 5.24) or push/twist-release action.

5.3.2   Selector Switches

The operator on a selector switch is rotated instead of pressed. Selector switches often use the same contact blocks as the push buttons. As the name implies, these switches allow the selection of one of the two or more circuits by rotating the operator to a certain position. For example, a selector switch can be used to select between the manual or automatic operation of an axis.

Figure 5.25 shows a three-position selector switch with an N.O. and an N.C. contact block. A *truth table* (or a target table) indicates which contacts are closed in each of the switch positions. Selector switches are available with the spring-return and/or maintained operator function, different truth tables, and in many styles and colors.

5.3.3   Indicator Lights

Indicator lights such as the one shown in Figure 5.26 are also called pilot lights. They are used to provide a visual indication of the machine status. They come in a variety of colors and shapes. There are also dual input push-to-test pilot lights where the light looks like a push

	Contacts	
Position	A	B
Left	X	0
Center	0	0
Right	0	X

X = Closed, 0 = Open.

(a)   (b)

Figure 5.25 Three-position selector switch. (a) Selector switch (Courtesy of *Rockwell Automation, Inc.*) [22]. (b) Truth table

Figure 5.26   Pilot light (Courtesy of *Rockwell Automation, Inc.*) [24]

button. Pressing on the light activates the test input to the light so that we can troubleshoot a potentially faulty bulb in the light.

5.4 Control Devices for AC Induction Motors

Three-phase AC induction motors require high currents and operate at high voltages. Therefore, across-the-line operation of a three-phase AC induction motor requires a control device called the *contactor*. In its most basic form, a contactor functions like a light switch found in a room on the wall. By turning the contactor switch on/off manually, we can start or stop a motor. The difference from an ordinary light switch is in the contacts. Since motors draw a significant amount of current, the contacts of the contactor are rated for higher currents.

Most motor applications involve starting or stopping a motor from a control circuit rather than by manually flipping a switch. Magnetic contactors are used in such applications. A *magnetic contactor* is like a relay in that it has a coil and a set of contacts. When the coil is energized by the control circuit, the contacts open (or close). These contacts are where the motor is wired and are rated for high currents.

When contactors are used in motor control applications, an additional component called *overload relay (OL)* is required. The OL protects the motor from damage due to excessive amounts of current. For example, if a conveyor driven by a motor gets jammed, the motor will continue to try to drive the belt and draw increasing amounts of current. When the excessive amount of current is drawn by the motor over a period of time, the OL will shut down the power to the motor.

Thermal OLs contain a heater element per phase of the motor. Motor phase currents go through the contactor and the OL terminals. Each motor phase current also goes through its heater element in the OL. If an overload condition occurs, the element heats up and causes a bimetallic component to trip the motor contacts. Figure 5.27 shows a contactor and an OL. The three-phase lines are shown by L1, L2, and L3. The contacts on the contactor are denoted by "M" and are shown as N.O. When the coil (not shown) of the contactor is energized, the "M" contacts close allowing each phase current to flow to the motor through the OL. If the contactor coil is de-energized, the "M" contacts open, which cuts the phase currents to the motor. The combination of a contactor and an overload relay is also called a *motor starter*. Commercial units that integrate the two components into a single package are available.

When a squirrel cage three-phase AC induction motor is started using across-the-line approach, it will draw a significant amount of current (about 600% of full-load current).

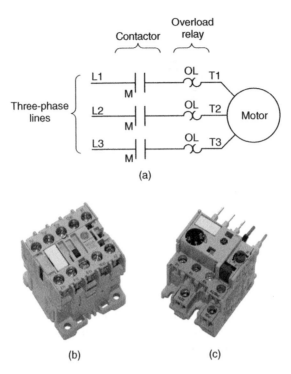

Figure 5.27   Contactor and overload relay schematic and commercial modules. (a) Schematic. (b) Commercial contactor unit [6]. (c) Commercial thermal overload relay unit [7]

This is not desirable. Various methods such as reduced-voltage starting, autotransformer, or wye-delta starters exist to address this issue. With the increasing availability of solid-state power devices, *soft starters* became popular. They limit the starting torque and current by gradually increasing the voltage over a user-programmed period of time. The same procedure can also be applied in stopping the motor.

5.4.1   Motor Control Circuit

A simple and common motor control circuit is shown in Figure 5.28. The operator interface consists of three buttons: E-STOP, STOP, and START. These buttons are mounted either on the door of a control panel or they can be mounted on a separate box. The diagram has two circuits: (1) control circuit and (2) power circuit. The control circuit was drawn using thin lines. The circuit shown with thick lines is the power circuit.

The power circuit consists of a magnetic contactor and an OL. The magnetic contactor is actuated by its CR coil. When the coil is energized, the contactor closes its contacts. The contacts can be opened by de-energizing the CR coil. The control circuit uses two of the main three-phase lines for its power. Often the voltage needed for the control circuit is much less (e.g., 24 VAC) than the voltage needed for the power circuit (e.g., 208 VAC). Therefore, a

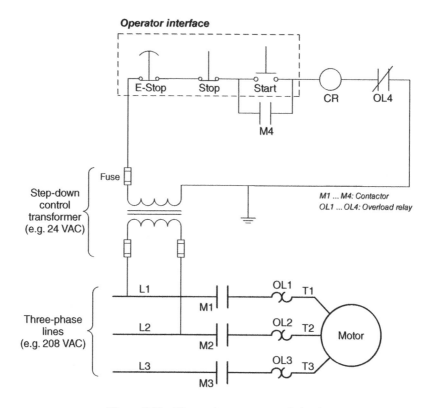

Figure 5.28 Three-wire motor control circuit

step-down control transformer is used to reduce the voltage. An N.C. momentary-type STOP push button, an N.O. momentary-type START push button, and an N.C. maintained-type (latching) E-STOP button are the pilot devices used to construct the operator interface. The N.C. OL4 contact used in the control circuit is an auxiliary contact on the overload relay. Similarly, the M4 contact in the control circuit is an auxiliary contact on the contactor.

If the E-STOP button was not engaged, pressing the START button energizes the CR coil of the magnetic contactor, which closes the M1, M2, and M3 contacts and starts the motor. When the coil is energized, the M4 contact also closes. The parallel branch formed by the M4 contact around the START button keeps the coil energized even when the START button is released. This is called *sealing in* the START button or the *holding circuit*.

If the STOP button is pressed, the power to the CR coil is interrupted, which opens all contacts of the contactor and stops the motor. Releasing the STOP button will not restart the motor. While the motor is running, if a condition causes an overload, the OL contacts will open. The OL1, OL2, and OL3 contacts will cut the power to the motor. The OL4 contact cuts the power in the control circuit. This way, when the overload thermal contacts cool down and close again, the motor cannot start suddenly by itself. If the E-STOP is pressed, it cuts the power to the control circuit and the contactor coil to stop the motor. Since the E-STOP button remains latched, the motor cannot be restarted by just pressing the START button. The E-STOP must be physically reset first by pulling the button up.

Problems

1. What is an optical encoder? Sketch the main components of an optical encoder and label them.
2. What creates the quadrature signal pattern in an incremental encoder? How can we determine the direction of rotation from the quadrature encoder signals?
3. Why does a 2000 line encoder create 8000 counts per revolution in quadrature decoding mode?
4. What is an index pulse? What is it used for in motion control?
5. Due to the low voltage outputs, SinCos encoders use complementary signals in each channel. Explain how the complementary channel approach may help improve noise immunity.
6. A servo motor has a 1024 line incremental encoder. If the motor is running at 3600 rpm and the controller uses quadrature decoding, how many counts per second will be generated by the encoder?
7. A machine axis uses a ball screw drive with 10 rev/in pitch. The axis motor is directly coupled to the ball screw. The axis must have ±0.0005 in accuracy. What is the minimum resolution the motor encoder should have?
8. An incremental encoder with 1024 PPR and quadrature decoding is used with a measuring wheel with 4 in diameter as shown in Figure 5.29. What is the minimum change in linear distance that can be measured with this wheel? If the measured part is moving at 4 ft/s speed, what will be the frequency of the encoder pulses in cts/s?

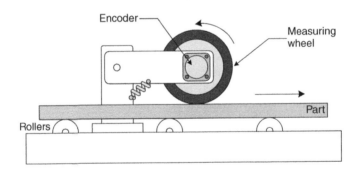

Figure 5.29   Measuring wheel for Problem 8

9. A SinCos encoder provides 512 sine waves per revolution.
   (a) If a 10-bit A/D is used by the motion controller circuitry, what will be the position resolution?
   (b) If the position counter reads 522 262 cts, what is the mechanical angular position of the encoder shaft?
10. A multiturn absolute encoder has a 4-bit main disk and three 3-bit disks for turn counts. What is the resolution of this encoder?
11. Compare the SSI® and EnDat® protocols based on the type of protocol, total number of wires and data transmission rates.

12. Consider a 512 line incremental encoder with quadrature decoder mounted on a motor. Assume that the controller has 2000 kHz sampling rate and uses the $1/T$ interpolation method with a 1 µs timer. What will be the percent velocity estimation error if a one-count error was made in the timer counts? What will be the percent velocity estimation error if the encoder is replaced with another one with 1024 PPR?

13. What are the three methods of detection used in photoelectric sensors?

14. Sketch the wiring diagram for a three-wire sinking sensor and a sourcing input card. Explain what happens when the sensor is triggered.

15. What is the purpose of using overload relays in AC induction motor control? How does a thermal overload relay work?

16. Explain how the control circuit given in Figure 5.28 works. What is the purpose of the OL4 contact in the circuit?

References

[1] An Engineering Guide to Position and Speed Feedback Devices for Variable Speed Drives and Servos. Technical Report, Emerson Industrial Automation, 2011.

[2] Cburnett (2006). Encoder disc (3-bit-binary).svg. http://en.wikipedia.org/wiki/File:Encoder_disc_(3-Bit _binary).svg (accessed 1 July 2012).

[3] Sabri Cetinkunt. *Mechatronics*. John Wiley & Sons, Inc., 2006.

[4] Danaher Industrial Controls (2003). Encoder Application Handbook. http://www.dynapar.com/uploadedFiles /_Site_Root/Service_and_Support/Danaher_Encoder_Handbook.pdf (accessed 28 April 2013).

[5] George Ellis. *Control Systems Design Guide*. Elsevier Academic Press, Third edition, 2004.

[6] Dmitry G. (2013) GE MC2A310AT1.JPG. http://commons.wikimedia.org/wiki/File:GE_MC2A310AT1.JPG (accessed 11 May 2013).

[7] Dmitry G. (2013) GE MTO3N.JPG. http://commons.wikimedia.org/wiki/File:GE_MTO3N.JPG (accessed 11 May 2013).

[8] Galil Motion Control, Inc. (2005) Application Note #1248: Interfacing to DB-28104 Sinusoidal Encoder Interface. http://www.galilmc.com/support/appnotes/econo/note1248.pdf (accessed 4 January 2013).

[9] Heidenhain (2011). EnDat 2.2 – Bidirectional Interface for Position Encoders. http://www.heidenhain.us /enews/stories_1012/EnDat.pdf (accessed 10 October 2014).

[10] Heidenhain, Corp. (2014) Incremental Linear Encoder MSA 770. http://www.heidenhain.us (accessed 10 October 2014).

[11] iC Haus (2013). BiSS Interface. http://www.ichaus.de/product/BiSS%20Interface, 2013. (accessed 27 January 2013).

[12] Jjbeard (2006). Encoder disc (3-bit).svg. http://commons.wikimedia.org/wiki/File:Encoder_Disc_(3-Bit).svg (accessed 1 July 2012).

[13] Lucasbosch (2014). SICK WL12G-3B2531 photoelectric reflex switch angled upright.png. http://commons .wikimedia.org/wiki/File:SICK_WL12G-3B2531_Photoelectric_reflex_switch_angled_upright.png (accessed 5 February 2013).

[14] Cory Mahn. Open vs. Closed Encoder Communication Protocols: How to Choose the Right Protocol for Your Application. Danaher Industrial Controls, 2005.

[15] Frank D. Petruzella. *Electric Motors and Control Systems*. McGraw-Hill Higher Education, 2010.

[16] W. Rebel (2012). Gray disc.png. http://commons.wikimedia.org/wiki/File:Gray_disc.png (2012) (accessed 1 July 2012).

[17] Rockwell Automation, Inc. (2014) Bul. 802MC – Corrosion-Resistant Prewired, Factory-Sealed Switches (2014). http://www.ab.com/en/epub/catalogs/3784140/10676228/4129858/6343016/10707215/print.html (accessed 24 November 2014).

[18] Rockwell Automation, Inc. (2014) Bul. 871TM Inductive Proximity Sensors, 3-Wire DC. http://www.ab.com /en/epub/catalogs/3784140/10676228/4129858/6331438/10706844/Bul-871TM-Inductive-Proximity-Sensors .html (accessed 24 November 2014).

[19] Rockwell Automation, Inc. (2014) Bul. 873P Analog or Discrete Output Ultrasonic Sensors (2014). http://ab.rockwellautomation.com/Sensors-Switches/Ultrasonic-Sensors/Analog-or-Discrete-Output-Ultrasonic-Sensors#overview (accessed 24 November 2014).

[20] Rockwell Automation, Inc. (2014) Bulletin 800F 22.5 mm Push Buttons, Flush Operator Cat. No. 800FP-F3 (2014). http://www.ab.com/en/epub/catalogs/12768/229240/229244/2531081/1734224/Momentary-Push-Button-Operators.html (accessed 24 November 2014).

[21] Rockwell Automation, Inc. (2014) Bulletin 800T, Emergency Stop Devices, 2-Position Metal Push-Pull Cat. No. 800TC-FXLE6D4S (2014). http://www.ab.com/en/epub/catalogs/12768/229240/229244/2531083/Emergency-Stop-Devices.html (accessed 24 November 2014).

[22] Rockwell Automation, Inc. (2014) Bulletin 800T/800H 3-Position Selector Switch Devices, Standard Knob Operator, Cat. No. 800T-J2A (2014). http://www.ab.com/en/epub/catalogs/12768/229240/229244/2531083/Selector-Switches.html (accessed 24 November 2014).

[23] Rockwell Automation, Inc. (2014) Bulletin 800T/800H 30.5 mm Push Buttons, Flush Head Unit, Cat. No. 800T-A1A (2014). http://www.ab.com/en/epub/catalogs/12768/229240/229244/2531083/Momentary-Contact-Push-Buttons.html (accessed 24 November 2014).

[24] Rockwell Automation, Inc. (2014) Bulletin 800T/800H Pilot Light Devices, Transformer Type Pilot Light, Cat. No. 800T-P16R (2014). http://www.ab.com/en/epub/catalogs/12768/229240/229244/2531083/Pilot-Light-Devices.html (accessed 24 November 2014).

[25] SICK (2012). The World of Motor Feedback Systems for Electric Drives. http://www.sick-automation.ru/images/File/pdf/Hyperface_e.pdf (accessed 25 March 2013).

[26] Siemens Industry, Inc. (2013) Basics of Control Components. http://www.industry.usa.siemens.com/services/us/en/industry-services/training/self-study-courses/quick-step-courses/downloads/Pages/downloads.aspx (accessed 29 March 2013).

[27] Texas Instruments (2008). TMS320x2833x, 2823x Enhanced Quadrature Encoder Pulse (eQEP) Module. http://www.ti.com/lit/ug/sprug05a/sprug05a.pdf (accessed 8 July 2014).

[28] US Digital Corp. (2014) E5 Optical Kit Encoder. http://www.usdigital.com (accessed 11 November 2014).

6

AC Drives

A drive amplifies small command signals generated by the controller to high-power voltage and current levels necessary to operate a motor. Therefore, the drive is also called an amplifier.

In motion control systems, each axis operates under closed-loop control. Typically, there are three loops to close in each axis, namely, current, velocity, and position loops. The motor's velocity and position are measured and fed back to the controller, while the motor's currents are measured and fed back to the drive. In other words, the drive closes the current loop. Yet, in recent trends, the line between the functions of a controller and the drive continues to blur. Many functions of the controller, including closing the velocity and position loops, are now done by the drive.

This chapter begins by presenting the building blocks of the drive electronics, namely, converter, DC link, and inverter. Control logic for the inverter using 120° and 180° conduction methods is discussed. The popular pulse-width modulation (PWM) control technique is explained. Then, basic closed-loop control structures implemented in the drive are introduced. Single-loop proportional-integral-derivative (PID) position control and cascaded velocity and position loops with feedforward control are explored in depth. Next, the implementation of the current loop using vector control for AC servo motors and induction motors is explained. Mathematical and simulation models of the controllers are provided. Control algorithms use gains that must be adjusted or tuned so that the servo system for each axis can follow its commanded trajectory as closely as possible. The chapter concludes by providing tuning procedures for the control algorithms presented earlier including practical ways to address integrator saturation.

6.1 Drive Electronics

The synchronous speed of the three-phase stator will be fixed if the stator is directly connected to three-phase AC power lines. To vary the speed, we need to adjust the voltage level and frequency of the three-phase sinusoidal waveforms supplied to the motor. The power electronics equipment designed for this purpose is called an *AC drive* or simply a *drive*.

Figure 6.1 shows the basic circuit blocks in a drive. AC drives convert AC power into DC and invert DC into variable voltage and frequency three-phase AC power. There are two types

Industrial Motion Control: Motor Selection, Drives, Controller Tuning, Applications, First Edition. Hakan Gürocak.
© 2016 John Wiley & Sons, Ltd. Published 2016 by John Wiley & Sons, Ltd.

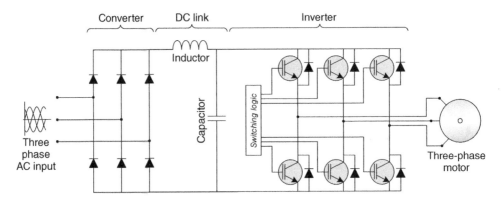

Figure 6.1 AC drive with pulse width modulation (PWM) inverter (Adapted by permission of Siemens Industry, Inc. [17])

of drives: *voltage source inverter (VSI)*-based drives and *current source inverter (CSI)*-based drives. The PWM switching technique is commonly used to achieve the transition from DC to AC in the inverter section. The PWM drives are energy efficient and provide high performance. The PWM is implemented in the *switching logic* circuit block of the drive. The VSI produces an adjustable three-phase PWM voltage waveform for the load (motor). The CSI outputs a PWM current waveform [18].

6.1.1 Converter and DC Link

A diode allows the current flow in one direction from positive to negative. If the current flow is reversed (negative to positive), the diode blocks the current flow. In a 60 Hz AC power source, voltage changes direction 60 times per second. If an AC source is connected to a diode, as shown in Figure 6.2a, it will have the effect of cutting the bottom part of the waveform. When the sine wave is in the positive half of the cycle, the diode conducts. When the sine wave is in the negative half of the cycle (bottom portion of the wave), the diode is reverse biased and, therefore, will not conduct. When the sine wave is again in its positive half of the cycle, the diode conducts. Essentially, the diode changes AC into DC (unidirectional). But, of course, this DC signal has very large ripple.

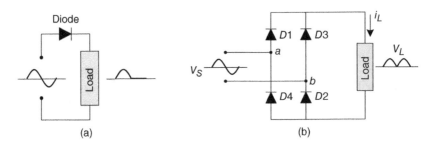

Figure 6.2 Converting AC into DC using diode rectifiers. (a) Diode has the effect of cutting the bottom part of the AC waveform. (b) Full-wave bridge rectifier

A *full-wave bridge rectifier* can be constructed as shown in Figure 6.2b. This circuit enables the current to exist in the negative half of the cycle. During the positive half of the input voltage, supply current enters the bridge from point *a*, flows through diode *D*1 and the load and returns through diode *D*2 and point *b* back to the supply. When the input wave is in its negative half cycle, the supply current enters the bridge from point *b*, flows through diode *D*3 and the load and returns through diode *D*4 and point *a* back to the supply. In the positive half of the wave, diodes *D*3 and *D*4 are reverse biased and do not conduct. Similarly, in the negative half of the wave, diodes *D*1 and *D*2 do not conduct. Note that the bridge rectifier steers the source current such that the load current i_L always flows through the load in the same direction. Hence, the alternating source current becomes unidirectional in the load and the negative half of the input voltage is folded up by the rectifier as seen in the load voltage (V_L) [9].

Figure 6.3a shows a *three-phase diode bridge rectifier* circuit. It converts the incoming three-phase AC line power into a DC voltage (with ripple). Therefore, this bridge is also called a *converter*. The diodes are numbered based on the sequence they conduct. The diode with the highest positive phase voltage out of *D*1, *D*3, and *D*5 will conduct. Similarly, the diode with the most negative phase voltage out of *D*2, *D*4, and *D*6 will conduct. Each diode conducts for 120° per cycle and a new diode begins to conduct after a 60° interval. The output waveform consists of portions of the line-to-line AC input voltage waveforms. This converter is also called a *six-pulse rectifier* since the waveform repeats itself in 60° intervals in one cycle. Three-phase bridge rectifiers are available in compact sizes as commercial products (Figure 6.3b).

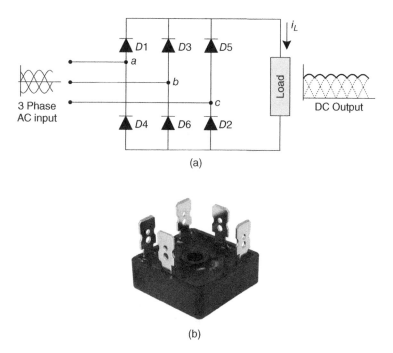

(a)

(b)

Figure 6.3 Three-phase converter circuit and the bridge rectifier as a commercial product. (a) Three-phase converter circuit. (b) Three-phase rectifier bridge. (Reproduced by permission of Vishay Intertechnology, Inc.) [1]

The *DC link* acts as a filter to smooth out the DC voltage output of the converter. The DC link section is designed differently depending on the drive type. In VSI-based drives, the DC link has a large capacitor. In CSI-based drives, the DC link has a large inductor. Sometimes, the VSI employes a small inductor (choke) in combination with the large capacitor as shown in Figure 6.4.

6.1.2   Inverter

An inverter creates AC output from a DC input using multiple switches. Switching devices such as bipolar transistors, MOSFETs, thyristors, or IGBTs (insulated gate bipolar transistor) are used in various inverters. The IGBT is the most common choice in many newer drives. An IGBT can turn ON/OFF in approximately 500 ns. The inverter shown in Figure 6.5a uses six high speed "switches" (IGBTs) that turn ON or OFF to create AC voltages for the phases of the motor from the DC bus voltage of the DC link. The diodes connected in parallel to each IGBT are called *free-wheeling diodes*. When a transistor is switched OFF, the inductor tries to force current through the transistor. The stored energy in the inductive load (motor) must be dissipated safely. The diode provides a bypass path for the current to protect the transistor [10]. Three-phase inverters are available in compact sizes as commercial products (Figure 6.5b).

6.1.2.1   Switching Logic-Conduction Method

There are two common methods used in controlling how the transistors are switched ON or OFF: 120° and 180° conduction. Figure 6.6 shows an inverter where the IGBTs are replaced with switches for simplification. The switching pattern determines the output frequency and the voltage from the inverter. If the line-to-neutral voltages for any of the phases were plotted, it can be seen that there are six discontinuities per cycle. They correspond to the switching points. Therefore, both of these conduction methods are often called the *six-step inverter*.

In the *120° conduction method*, the switching pattern is arranged so that one of the three legs of the inverter is connected to the positive DC bus terminal, another to the negative DC bus terminal, and the third leg is disconnected. For example, in Figure 6.6, the first leg containing $T1$ and $T4$ can be connected to the positive terminal by closing $T1$ (turning transistor on) and opening $T4$ (turning transistor OFF). The second leg containing $T2$ and $T5$ is connected to the negative terminal by opening $T2$ and closing $T5$. Finally, the third leg is disconnected by keeping both $T3$ and $T6$ open. This switching pattern would allow current to flow from the

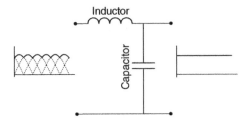

Figure 6.4   DC link smooths out the resulting DC bus voltage

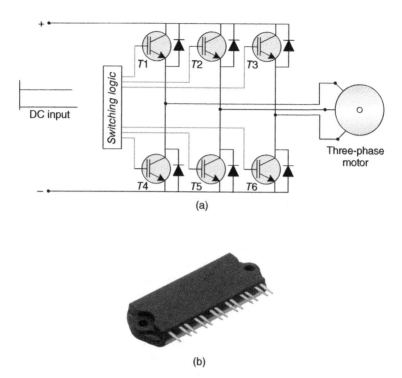

(a)

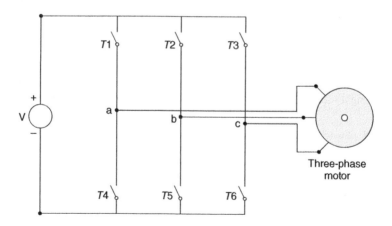

(b)

Figure 6.5   Three-phase inverter circuit and a commercial inverter module. (a) Inverter circuit with six IGBTs (*T*1 through *T*6) (Adapted by permission of Siemens Industry, Inc.) [17]. (b) Three-phase inverter. (Reproduced by permission of Vishay Intertechnology, Inc.) [2]

Figure 6.6   Simplified inverter with switches replacing the IGBTs

positive side of the DC bus through $T1$, phase A and B windings in the motor back through $T5$ into the negative terminal of the DC bus. Notice that in this method, a switch is connected to the positive and negative terminals of the DC bus for 120° intervals in one cycle of operation, hence the name of the method.

When the inverter operating with the 120° conduction method is connected to a brushless motor, we obtain the rotation as shown in Figure 4.7. Switching patterns are shown in Table 6.1 where each row corresponds to one of the diagrams in Figure 4.7. Again, note that in each step, only two switches are on (transistors conducting). As a result of this, in each step, two of the stator phases are connected to the DC bus, while the third one is left floating.

In the *180° conduction method*, each switch is closed for 180° of the output cycle and opened for the rest of the 180°. The timing of the individual switches is adjusted such that 120° after closing $T1$, $T2$ is closed. Similarly, 120° after closing $T2$, $T3$ is closed. Using this technique, an approximate sinusoidal waveform can be generated. The frequency of the resulting sinusoidal wave depends on how fast the switches are opened and closed (switching frequency).

In either of the conduction methods, the magnetic field rotates 60° in each step. But depending on the conduction method, the resulting phase voltage waveforms are different. When driving an inductive load, such as a motor, the line-to-line waveform is independent of the load characteristics with the 180° conduction method, whereas in the 120° case, the waveforms will be affected by the transient current behavior [11].

Table 6.2 shows the switching patterns in this mode. Notice that one switch closes and another opens every 60°. Also, in each step there are three switches closed. Unlike the 120° case, there are no floating terminals. Each motor terminal is connected to voltage. As a result, the supply current is apportioned among the phases and current flows in each phase.

6.1.2.2 Pulse-Width Modulation (PWM) Control

AC drives can vary the speed of the motor by adjusting the voltage and frequency of the three-phase sinusoidal waveforms supplied to the motor. When a fixed DC bus voltage is used with an uncontrolled rectifier bridge as discussed so far, we need a way to vary the amplitude of the voltage waveforms generated by the inverter.

PWM is a very efficient way of providing variable voltage levels. There are many PWM techniques including sinusoidal, space-vector, and hysteresis band current control [3]. By varying how long a square wave stays on in a given period, a variable effective voltage output

Table 6.1 Switching patterns for 120° conduction method

Switching interval (°)	Hall sensor	Transistors						Terminal voltages		
		$T1$	$T2$	$T3$	$T4$	$T5$	$T6$	A	B	C
0–60	001	OFF	OFF	ON	OFF	ON	OFF	f[a]	−	+
60–120	011	ON	OFF	OFF	OFF	ON	OFF	+	−	f
120–180	010	ON	OFF	OFF	OFF	OFF	ON	+	f	−
180–240	110	OFF	ON	OFF	OFF	OFF	ON	f	+	−
240–300	100	OFF	ON	OFF	ON	OFF	OFF	−	+	f
300–360	101	OFF	OFF	ON	ON	OFF	OFF	−	f	+

[a]Motor terminal is disconnected. Terminal voltage is floating.

Table 6.2 Switching patterns for 180° conduction method

Switching interval (°)	Hall sensor	Transistors						Terminal voltages		
		T1	T2	T3	T4	T5	T6	A	B	C
0–60	001	ON	OFF	ON	OFF	ON	OFF	+	−	+
60–120	011	ON	OFF	OFF	OFF	ON	ON	+	−	−
120–180	010	ON	ON	OFF	OFF	OFF	ON	+	+	−
180–240	110	OFF	ON	OFF	ON	OFF	ON	−	+	−
240–300	100	OFF	ON	ON	ON	OFF	OFF	−	+	+
300–360	101	OFF	OFF	ON	ON	ON	OFF	−	−	+

can be obtained. *Duty cycle* is defined as the percent ratio of the on-time to the period of the square wave (Figure 6.7a):

$$D = \frac{t_{\text{on}}}{T} \times 100$$

where t_{on} and T are the on-time and the period of the wave, respectively. D is the percent duty cycle. The average voltage level can then be found from

$$V_{\text{ave}} = D V_H + (1 - D)V_L$$

Usually V_L is taken as zero. Therefore, the average voltage is directly proportional to the duty cycle (Figure 6.7b). Since the signal toggles between the ON and OFF states very fast, the load does not get affected by the switching and only sees the average voltage level.

To generate a PWM signal, a reference sinusoidal signal, v_{ref}, and a triangular carrier signal, v_{carrier}, are used as control signals as shown in Figure 6.8. The voltage and frequency of the carrier signal are fixed, whereas the reference signal can have adjustable voltage and frequency. At any given instant, the magnitudes of the signals are compared to each other. If the magnitude of the reference signal is greater than the carrier, then the PWM signal is set to 1 (switch closed). Otherwise, it is set to zero (switch open).

The PWM signal is used to trigger the transistors in each leg of the inverter. Let us look at the first leg of the simplified inverter in Figure 6.6 where transistors $T1$ and $T4$ switch power to phase A. If $T1$ is closed and $T4$ is open, the DC bus voltage V will be applied to phase A. If, on the other hand, $T1$ is open and $T4$ is closed, the phase A voltage will be zero. We can control the switching of the transistors in each leg based on the difference between the reference and the carrier signals [6]

$$\Delta v = v_{\text{ref}} - v_{\text{carrier}}$$

Then, the switching conditions for phase A are

$$\Delta v_a > 0, \quad T1 \text{ is closed and } T4 \text{ is open}$$

$$\Delta v_a < 0, \quad T4 \text{ is closed and } T1 \text{ is open}$$

By turning the inverter switches ON/OFF with varying duty cycles, the PWM signal allows us to apply a *variable* phase voltage. The control signals of the other two legs are sinusoidals that are shifted 120° with respect to each other for a balanced three-phase system. However,

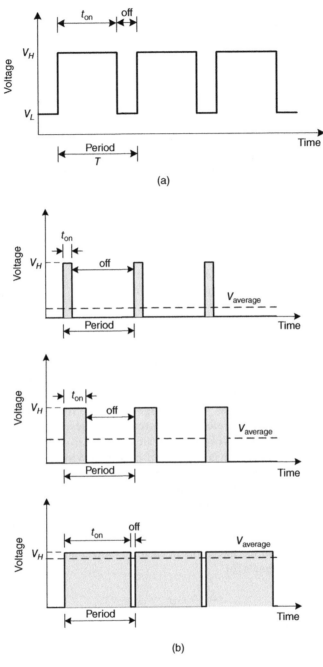

(a)

(b)

Figure 6.7 Pulse-width modulation (PWM). (a) Duty cycle in pulse width modulation (PWM). (b) Average voltage output is proportional to the duty cycle

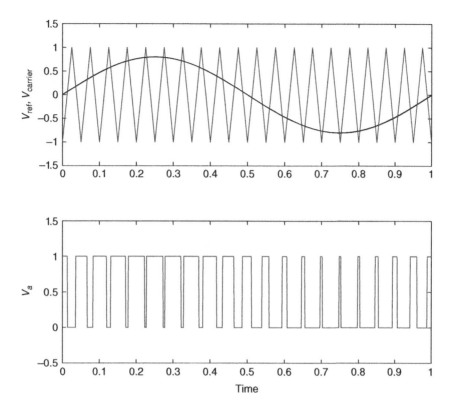

Figure 6.8   Triangular carrier signal and sinusoidal reference signal to create a corresponding PWM signal

the same carrier signal is used for all phases. Figure 6.9 shows two phase reference signals and the resulting PWM signals for phase A and phase B legs of the inverter.

Figure 6.10 shows the line-to-line voltage V_{ab} applied to the motor by the "a" and "b" legs of the inverter. This voltage can be found from the difference of the V_a and V_b PWM signals, as shown in Figure 6.9. The signal has a dominant sinusoidal component at the same frequency as the reference sinusoidal frequency. Its amplitude is given by the $V_{ref}/V_{carrier}$ ratio. Hence, by changing the frequency and amplitude of the reference signal, we can control the frequency and amplitude of the line-to-line voltages applied to the motor [5, 19].

PWM signals tend to be in the range of 2–30 kHz. The duty cycle of the PWM signal defines the average voltage applied to the stator, which is proportional to the motor speed. If the duty cycle is increased, the motor speed will increase. In addition to the efficiency advantage, the PWM method also allows easy adjustment of the DC bus voltage. If the DC bus voltage of the controller is higher than the rated voltage of the motor, the magnitude of the PWM signal can easily be adjusted in software to match the rated voltage of the motor. As a result, the controller can be interfaced to motors with different rated voltages.

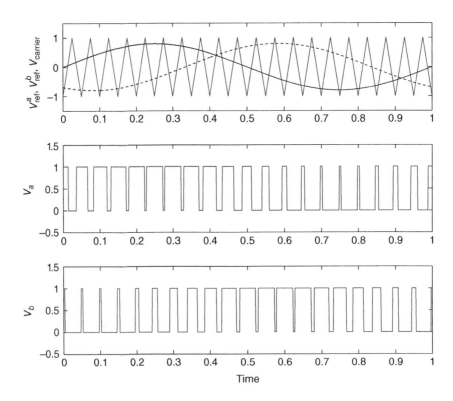

Figure 6.9 Two-phase sinusoidal reference signals shifted 120° and the corresponding PWM signals

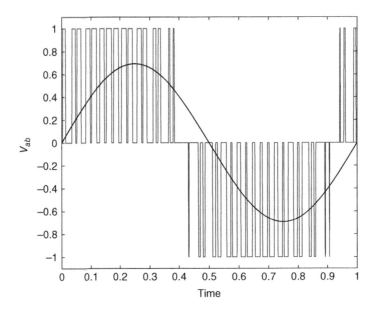

Figure 6.10 PWM signal for the line-to-line V_{ab} voltage

6.1.2.3   Simulation Model of a PWM Inverter

The model of the PWM inverter was built hierarchically. Figure 6.11a shows the top-level model. It requires three reference voltages as inputs corresponding to the desired phase currents i_a, i_b, i_c. The model generates the PWM phase voltages V_a, V_b, V_c with respect to the DC bus center tap.

When the top-level model is opened, the next level of detail can be seen as shown in Figure 6.11b. The model requires DC bus voltage, carrier frequency, and amplitude to be specified by the user. This information is then used to build the triangular carrier signal and the bus voltage in the model. The carrier signal is subtracted from the desired phase current

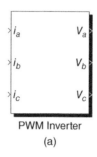

PWM Inverter

(a)

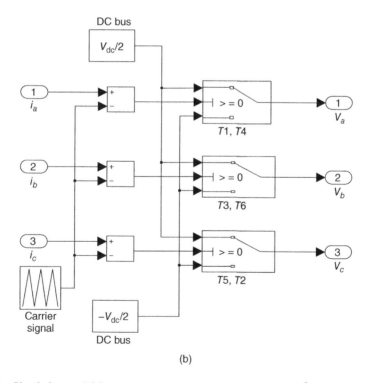

(b)

Figure 6.11   Simulation model for the PWM inverter. (a) Top-level Simulink® model for PWM inverter. (b) Details of the PWM inverter model

references. Each of the three switch blocks corresponds to a transistor pair in Figure 6.6. If the current difference (middle input of a switch block) is positive, $V_{dc}/2$ is supplied as the phase voltage; otherwise, $-V_{dc}/2$ becomes the phase voltage.

6.2   Basic Control Structures

Motion control requires precise control of velocity and position. Although several control structures are possible, there are three commonly used control structures:

1. Cascaded velocity and position loops
2. Single-loop PID position control, and
3. Cascaded loops with feedforward control.

6.2.1   Cascaded Velocity and Position Loops

This structure is by far the most common and contains a velocity loop which is enclosed in a position loop. These are also referred to as the *outer loops*. The torque is regulated by the *inner loop* by controlling the currents to the motor. We will explore it further in Section 6.3.

In the past, motion control systems often used a velocity sensor, such as a tachometer, and a position sensor such as an encoder. Today most motion control systems use just an encoder to measure the actual motor shaft position and derive the velocity from it using a software algorithm. This is represented by the $\frac{d}{dt}$ block. Typically, a PI controller is the choice in the velocity loop as shown in Figure 6.13. The proportional gain ("P") enables high-frequency response and the stability of the loop, while the integral component ("I") drives the error to zero [8]. The velocity loop enables the system to follow rapidly changing velocity commands and also provides resistance to load disturbances with high frequencies.

The controller gains of these cascaded loops are adjusted (tuned) starting from the inner most loop and working outward. After the inner loop and the velocity loop are tuned, the position loop looks like Figure 6.14 where the velocity loop acts just as another dynamic component (a low-pass filter). The choice of the controller for the position loop can be a "P" or PI controller. The performance of each loop affects the performance of the next loop enclosing it. Therefore, each loop must be tuned to be as responsive as possible as we work outward in this process. The tuning process is explained in Section 6.5.

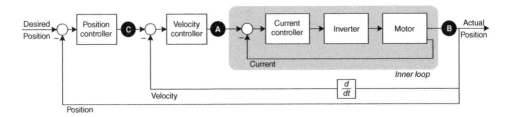

Figure 6.12   Cascaded velocity and position loops of the control system for an axis

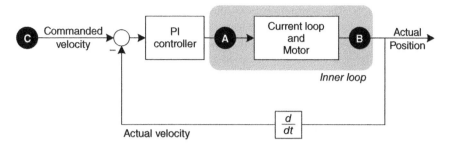

Figure 6.13   Velocity loop

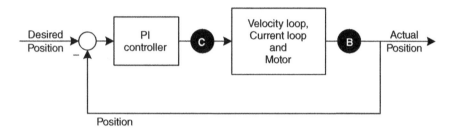

Figure 6.14   Position loop

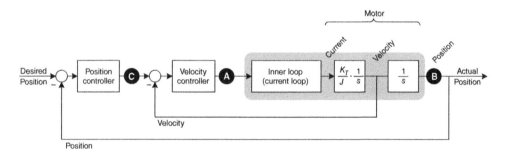

Figure 6.15   Double integrators

6.2.1.1   Simplified Theoretical Model for the Velocity Loop

An important difference of a motion control system from many of the other control systems is that it contains *double integrators* ($\frac{1}{s^2}$). In Figure 6.15, the double integrators were split into two single integrators in series. The first integration gets us from current to velocity which is then integrated again to get to the shaft position $\theta(s)$. Conceptually, the velocity loop encloses one of the integrators, while the position loop encloses the second integrator.

There is a vast amount of theory on control systems. The reader is referred to Appendix A for an overview. Here, we will make some assumptions to simplify the cascaded system and apply theoretical tools. These simplifications allow us to gain further insight into the inner workings of the control system and predict its performance.

In its simplest form, we can ignore all friction losses and lump the motor inertia and reflected load inertia into J. Then, the following equation governs the motor response

$$T = J\ddot{\theta}$$

where torque T is proportional to the current ($T = K_T i$). The simplification here is to model the motor electrical components as a gain K_T (motor torque constant). Recall that the torque produced by an AC motor is proportional to the quadrature current component which is what the inner loop regulates. Taking Laplace transform of both sides and rearranging it in transfer function format gives

$$\frac{\theta(s)}{I(s)} = \frac{K_T}{J}\frac{1}{s^2}$$

Note that one of these double integrals is to go from velocity to position. Then, we can rewrite the same equation as

$$\frac{s\,\theta(s)}{I(s)} = \frac{K_T}{J}\frac{1}{s}$$

But $s\,\theta(s) = \omega(s)$. In other words, the derivative of position $\theta(s)$ in the Laplace domain is the velocity $\omega(s)$ in the Laplace domain. Then,

$$\frac{\omega(s)}{I(s)} = \frac{K_T}{J}\frac{1}{s}$$

Putting it all together, we can arrive at the simplified velocity controller block diagram as shown in Figure 6.16.

The closed-loop transfer function for the velocity controller is given by [13]

$$\frac{\omega(s)}{\omega_d(s)} = \frac{G(s)}{1 + G(s)} \tag{6.1}$$

where $G(s)$ is the gain in the forward path of the system and is equal to the multiplication of all the gains

$$G(s) = \left(K_p + \frac{K_i}{s}\right)K_c\frac{K_T}{Js} \tag{6.2}$$

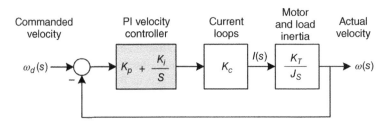

Figure 6.16   Simplified theoretical velocity controller block diagram

Here, K_p is the *proportional gain* and K_i is the *integral gain* of the PI controller. After substituting Equation (6.2) into (6.1) and simplifying, the closed-loop transfer function can be found as

$$\frac{\omega(s)}{\omega_d(s)} = \frac{K_c K_T K_p s + K_i}{Js^2 + K_c K_T K_p s + K_c K_T K_i}$$

For a given motor, K_T is fixed. After tuning the current controller with the motor, K_C is also fixed. Therefore, we can combine the K_C and K_T gains into a single gain $K_m = K_c K_T$ and simplify the transfer function further as

$$\frac{\omega(s)}{\omega_d(s)} = \frac{K_m K_p s + K_i}{Js^2 + K_m K_p s + K_m K_i} \tag{6.3}$$

As can be seen, the system response is only a function of the velocity controller gains K_p and K_i once the inertia is determined by the design of the system.

Equation (6.3) can be modified into

$$\frac{\omega(s)}{\omega_d(s)} = \frac{\frac{1}{J}(K_m K_p s + K_i)}{s^2 + \left(\frac{K_m K_p}{J}\right)s + \left(\frac{K_m K_i}{J}\right)} \tag{6.4}$$

The standard form for a second-order system is defined as [13]

$$s^2 + 2\zeta\omega_n s + \omega_n^2$$

If we match the coefficients of the denominator of Equation (6.4) to the coefficients of the standard form, we can obtain the natural frequency ω_n and damping ratio ζ of the closed-loop, system in terms of the controller gains K_p and K_i as

$$\omega_n = \sqrt{\frac{K_m K_i}{J}} \tag{6.5}$$

$$\zeta = \frac{K_m K_p}{2\omega_n J} \tag{6.6}$$

As can be seen from Equation (6.5), the integral gain K_i directly influences the natural frequency of the closed-loop system. The proportional gain K_p relates the natural frequency to the bandwidth. If a desired natural frequency is specified, the integral gain can be found from Equation (6.5) as

$$K_i = \frac{J\omega_n^2}{K_m}$$

The damping ratio must be selected based on the specific requirements of the application. However, in motion control systems, a damping ratio of $\zeta = 1$ is often selected to try to obtain a critically damped response. Then, using $\zeta = 1$ in Equation (6.6), the proportional gain can be calculated as

$$K_p = 2\sqrt{\frac{JK_i}{K_m}} \tag{6.7}$$

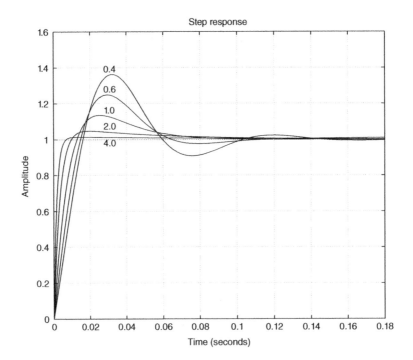

Figure 6.17   Closed-loop response of the velocity loop for various damping ratios

If K_p is varied while keeping K_i at a set value, the damping ratio will change. In Figure 6.17, the closed-loop response of the velocity control loop is shown for various values of damping ratios. As the damping ratio goes below 1.0, the response becomes more oscillatory with significant overshoot. At $\zeta = 1.0$, the response should theoretically have no overshoot but the zero ("s" term) in the numerator in Equation (6.4) influences it to cause an overshoot of about 13%. In general, an overshoot of 10–15% is acceptable in the velocity loop.

Vendors use internal and product-specific units for the gain settings in their products. There-fore, it is often hard to correlate the gains computed from theoretical analysis to the actual settings needed for a particular commercial product. In addition, often it is very difficult to obtain an accurate mathematical model of the actual axis being tuned. Sophisticated methods and instruments, such as dynamic signal analyzers (DSA), are available for more accurate system modeling. However, most often the controller is tuned by applying simple tuning rules while working with the actual machine. Also, some motion controllers are sophisticated enough to do automatic tuning, which often is a good starting point for further manual fine tuning of the gains.

6.2.2   Single-Loop PID Position Control

This control structure, as shown in Figure 6.18, does not have a velocity loop. It consists of only the position loop. The position loop encloses double integrators (the two $1/s$ terms in Figure 6.18). This is the fundamental difference of a PID *position* controller from other

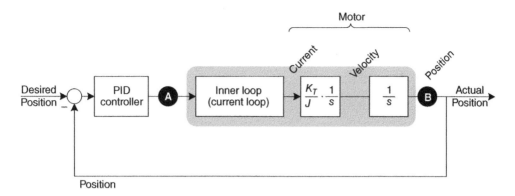

Figure 6.18   Single-loop PID position control encloses double integrators

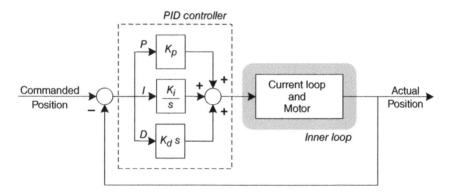

Figure 6.19   PID controller has three gains

applications of a PID controller where the system under control is most likely a single integrator [8]. The derivative ("D") signal is added to the controller to deal with the phase lag of the double integrators.

The PID controller has a proportional gain K_p, an integral gain K_i, and a derivative gain K_d to be tuned as shown in Figure 6.19. Each control signal plays a certain primary role in a closed-loop control system. But it should be kept in mind that these signals do not function completely independently from each other. They tend to interact with each other. Let us look at the individual roles of the K_p, K_i, and K_d gains in a PID position controller.

6.2.2.1   The Role of the K_p Gain

Assume that we have a linear axis with a carriage and a ball screw as discussed in Section 3.3.3 in Chapter 3. It is also assumed that the friction, motor, and screw inertia are negligible. The motor is interfaced to a motion controller with the closed-loop control system, as shown in Figure 6.20. From Equation (3.24) in Chapter 3, the reflected load inertia is $J = m/(\eta N^2)$.

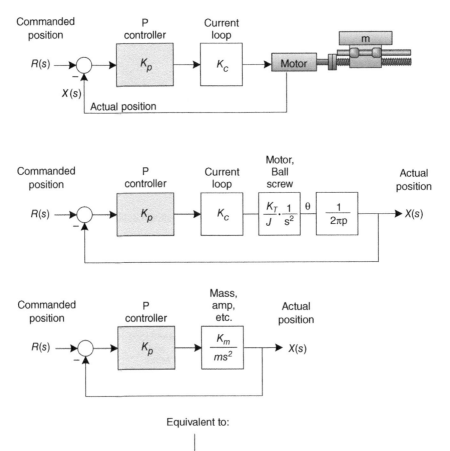

Figure 6.20   Motion controller with P control connected to a linear axis ($K_m = 1$)

The angular position of the motor shaft is converted into the linear position of the carriage by multiplying by the transfer function $1/(2\pi p)$, where p is the pitch of the ball screw. The overall transfer function for the current loop and the linear axis is then given by

$$G(s) = \frac{K_c K_T \eta N^2}{2\pi p m s^2}$$

$$= \frac{K_m}{ms^2}$$

where $K_m = (K_c K_T \eta N^2)/(2\pi p)$. To simplify the notation, let $K_m = 1$.

Then, the closed-loop transfer function (CLTF) of this system is

$$\frac{X(s)}{R(s)} = \frac{K_p}{ms^2 + K_p} \tag{6.8}$$

This transfer function tells us how the closed-loop system, including the motion controller and the linear axis, converts its commanded position input $R(s)$ into the carriage position output $X(s)$.

Now, let us examine the dynamic behavior of an ideal spring–mass system, as shown in Figure 6.21. This is an ideal system where there is no friction or any other energy losses. A force $f(t)$ is applied on the mass, which stretches the spring. At the instant when the displacement is equal to $x(t)$, the free-body diagram is shown in Figure 6.21. Using the free-body diagram, we can apply Newton's second law to derive the equation of motion as follows:

$$\sum F = m\ddot{x}$$
$$f(t) - kx = m\ddot{x}$$

If we rearrange this equation and take its Laplace transform, we can obtain

$$(ms^2 + k)X(s) = F(s)$$

Then, the transfer function for the system is

$$\frac{X(s)}{F(s)} = \frac{1}{ms^2 + k} \tag{6.9}$$

The denominator of a transfer function governs the dynamic behavior of the system. Comparing the denominator of Equations (6.8) and (6.9) reveals that the proportional gain K_p

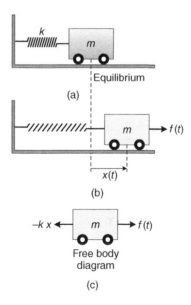

(a)

Equilibrium

(b)

$x(t)$

$-kx \leftarrow \boxed{m} \rightarrow f(t)$

Free body diagram

(c)

Figure 6.21 (a) Spring–mass system at rest; (b) displaced through distance $x(t)$ due to force $f(t)$; (c) free-body diagram

in the controller has the same role as the spring constant k in the simple spring–mass system, as shown in Figure 6.21. In other words, the proportional gain adds a *virtual spring* to the linear axis system. The higher the K_p is set, the stiffer the virtual spring will be in the system.

If the ideal simple spring–mass system is set in motion by the force $f(t)$, then it will oscillate back and forth. Likewise, if the commanded position, $r(t)$, for the linear axis is a step change, then the carriage will oscillate back and forth as shown in Figure 6.20, while the controller tries to settle it at the commanded position. Essentially, the closed-loop system made up of the linear axis (with no friction) and the controller will exhibit the same behavior as an oscillating simple spring–mass system.

6.2.2.2 The Role of the K_d Gain

Let us assume that the same linear axis with the motion controller in Figure 6.20 is set up to have a PD controller as shown in Figure 6.22. Again, let $K_m = 1$ for simplification. Then, the

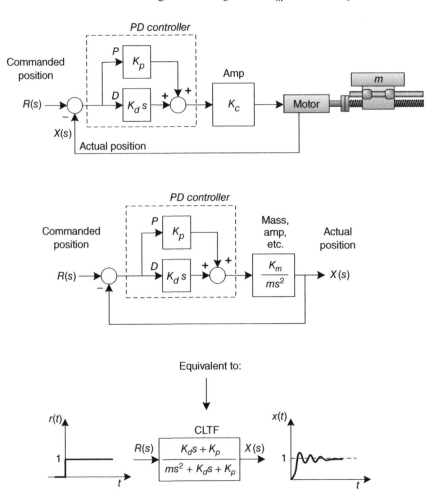

Figure 6.22 Motion controller with PD control connected to a linear axis ($K_m = 1$)

CLTF of this system is

$$\frac{X(s)}{R(s)} = \frac{K_d s + K_p}{ms^2 + K_d s + K_p} \tag{6.10}$$

Let us now add a damper to the simple system in Figure 6.21 to create the spring–mass–damper system in Figure 6.23. A damper applies a force that is proportional to the speed of its end point ($F_d = b\dot{x}$).

The effect of damping is to dissipate the energy in the system which will ultimately bring the mass to rest. Using the free-body diagram, we can apply Newton's second law to derive the equation of motion as follows

$$\sum F = m\ddot{x}$$
$$f(t) - kx - b\dot{x} = m\ddot{x} \tag{6.11}$$

If we rearrange Equation (6.11) and take its Laplace transform, we can obtain the following transfer function

$$\frac{X(s)}{F(s)} = \frac{1}{ms^2 + bs + k} \tag{6.12}$$

If we compare the denominator of Equation (6.10) to (6.12), we can see that the derivative gain K_d plays the same role as the damping coefficient b in the spring–mass–damper system in Figure 6.23. Note that the linear axis does not have a physical spring or damper. Yet, if it is connected to a motion controller with PD control, it acts like a simple spring–mass–damper system when a step position command is sent to it. The proportional gain adds a *virtual spring* and the derivative gain adds a *virtual damper* to the linear axis system. The higher the K_p is set, the stiffer the virtual spring will be in the system. The higher the K_d gain is, the faster the oscillations will die out.

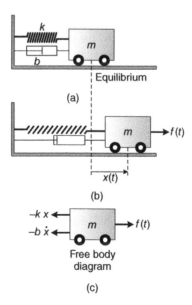

Figure 6.23 (a) Spring–mass–damper system at rest; (b) displaced through distance $x(t)$ due to force $f(t)$; (c) free-body diagram

The spring–mass–damper system will oscillate and settle about the equilibrium point once it is set in motion by the force $f(t)$. The closed-loop system with the linear axis (with no friction) under the control of the PD controller exhibits the same behavior when it receives a step change in its position command.

6.2.2.3 The Role of the K_i Gain

Let us assume that the same linear axis with the motion controller in Figure 6.22 is now set up to have a PID controller as shown in Figure 6.24. The CLTF for this system is ($K_m = 1$)

$$\frac{X(s)}{R(s)} = \frac{K_d s^2 + K_p s + K_i}{ms^3 + K_d s^2 + K_p s + K_i}$$

If we compare the denominator of this Equation to (6.12), we see that they do not match. In fact, unlike the cases of the K_p and K_d gains, there is no physical device that corresponds to the K_i gain in the simple spring–mass–damper system. Then, what does the K_i gain do?

The integral gain K_i is effective in countering steady-state errors. For example, positioning errors around the target position due to mechanical friction can be compensated by using K_i. Another situation is when a dead weight is applied to the servo axis after it is in position. If there is no K_i gain, the constant torque applied by the weight will cause the motor shaft to rotate slightly out of position. The weight will be held by the K_p gain since the increased position error multiplied by this gain will create more motor torque. The K_d gain will be inactive since there is no speed. Consequently, the weight will be supported by the motor, but there will be a static position error. The integrator creates a control signal that is proportional to the accumulation of error over time. In other words, the integrator signal is equal to the area under the error-versus-time curve of the system. The constant error caused by the dead weight will accumulate in the integrator over time. This will contribute an additional command signal to the amplifier, resulting in additional torque beyond the torque due to the K_p gain. As a result, not only the weight will be supported but the motor shaft will also move back to the target position making the position error zero. The K_i gain determines how fast the system can eliminate the position error (also called following error) and go back to being in position when subjected to such disturbances.

Figure 6.25 shows the block diagram of a system where a vertical load attached to a pulley is supported by the motor under PID position control.

As before, let $K_c = K_T = 1$ for simplification. The output of the system can be written as

$$\theta(s) = G_{\text{PID}}(s)G_s(s)E(s) - G_s(s)D(s) \tag{6.13}$$

Also,

$$E(s) = R(s) - \theta(s) \tag{6.14}$$

Solving for $\theta(s)$ from Equation (6.14) and substituting into (6.13) gives

$$E(s) = \frac{1}{1 + G_{\text{PID}}(s)G_s(s)}R(s) + \frac{G_s(s)}{1 + G_{\text{PID}}(s)G_s(s)}D(s)$$

Here, the first term is the position error due to the input command and the second term is the position error due to the disturbance on the system. The disturbance caused by the weight can

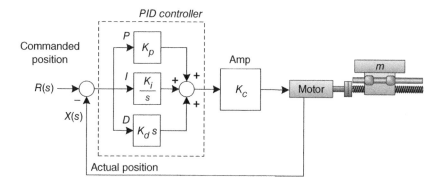

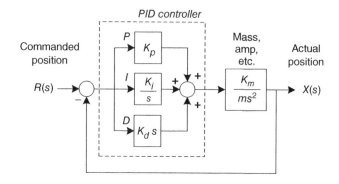

Equivalent to:

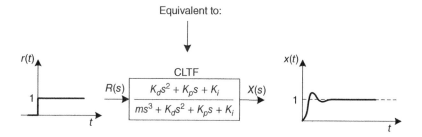

Figure 6.24 Motion controller with PID control connected to a linear axis ($K_m = 1$)

be modeled as a step disturbance after the motor is in position. To simplify the equations, let us assume that the magnitude of the disturbance torque is unity ($mgr = 1$). Hence, $D(s) = \frac{1}{s}$.

We need the final error due to the disturbance ($e_D(\infty)$) to be zero so that the system can compensate for the weight and still be in position. Using the final value theorem [13]

$$e_D(\infty) = \lim_{s \to 0} sE_D(s)$$

$$= \frac{s\, G_s(s)}{1 + G_{PID}(s)G_s(s)} D(s) \tag{6.15}$$

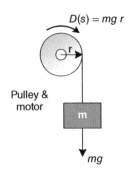

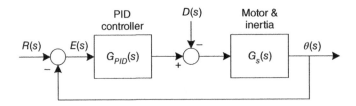

Figure 6.25   Motor with vertical load under PID position control

where

$$G_{\text{PID}}(s) = K_p + \frac{K_i}{s} + K_d s$$

and

$$G_s(s) = \frac{1}{Js^2}$$

with J as the total motor and pulley inertia. Substituting all of these transfer functions into (6.15) gives

$$e_D(\infty) = \lim_{s \to 0} \left(\frac{1}{Js^2 + K_d s + K_p + \frac{K_i}{s}} \right) \tag{6.16}$$

As the limit ($s \to 0$) is applied, the $\frac{K_i}{s}$ term goes to infinity. Consequently, the error due to the weight disturbance goes to zero. Hence, the integral gain K_i enables the system to reject constant disturbances.

If the controller did not have a K_i gain (just PD controller), then Equation (6.16) would become

$$e_D(\infty) = \lim_{s \to 0} \left(\frac{1}{Js^2 + K_d s + K_p} \right)$$

In this case, as the limit ($s \to 0$) is applied, we would get a nonzero finite error

$$e_D(\infty) = \frac{1}{K_p}$$

The error can be reduced by increasing the K_p gain, but this would also lead to more oscillations and overshoot.

6.2.3   Cascaded Loops with Feedforward Control

Servo systems need high gains for good performance with rapid, stable, accurate response, and ability to reject disturbances. However, high gains can cause instability, especially the integral gain in a PID controller. One solution is to use the integrator selectively only when in position as shown later in Example 6.5.2. But this time following errors occur *during* the motion even though the final following error is taken out by the integrator once the motion stops. Eliminating following errors during the motion is important in multi-axis contour following applications. In such applications, the following error in one axis can adversely affect the performance of the other axes and the overall machine performance. Then, how do we reduce or eliminate the following error during motion?

A popular approach in industrial motion controllers is the use of feedforward control along with the traditional feedback control based on the PID controller. Feedforward gains work outside the traditional feedback loops and as a result they do not lead to instability. In the typical feedback control approach, as shown in Figure 6.12, a velocity command (motion) is given to the motor only after a position following error occurs. In other words, the system takes an action only after the axis has already fallen behind in the desired trajectory. In the velocity feedforward approach, the required velocity to follow the desired trajectory is calculated and injected directly into the velocity loop without having to wait for an error to occur first. Such use of feedforward and feedback gains together can significantly improve the system performance.

6.2.3.1   Changing the PID Controller Structure

We will continue to study the traditional PID structure, as shown in Figure 6.19, but will first make a slight modification. In real systems, the mechanism will exhibit some friction which is often modeled as viscous damping in mathematical models. Let us assume that the energy dissipation caused by the friction in the mechanism can be modeled by adding a damping term to our previous model with double integrators $\left(\frac{K_m}{ms^2} \right)$. Then, the system transfer function can be rewritten as

$$G(s) = \frac{K_m}{ms^2 + bs}$$
$$= \frac{K_m}{s(ms + b)} \tag{6.17}$$

where b is the viscous damping coefficient representing the energy dissipation characteristic of the mechanism. Note that there is no physical damping in the system. Instead, this term was added to account for the fact that friction dissipates energy and would eventually slow down the motion of the system.

The transfer function for the PID controller is

$$C(s) = K_p + \frac{K_i}{s} + K_d s \tag{6.18}$$

Given these two transfer functions and the system configuration in Figure 6.26, the CLTF can be found from

$$CLTF = \frac{C(s)G(s)}{1 + C(s)G(s)} \tag{6.19}$$

Figure 6.26 Control structure with transfer functions for the PID position controller and the system

After substituting Equations (6.17) and (6.18) into (6.19) and some algebra, we can obtain

$$\text{CLTF} = \frac{K_m(K_d s^2 + K_p s + K_i)}{ms^3 + (b + K_m K_d)s^2 + K_m K_p s + K_m K_i} \tag{6.20}$$

Hence, the traditional PID structure creates a second-order polynomial in the numerator of the system CLTF. This polynomial is a function of the controller gains. It has two zeros (roots). Depending on the choice of the controller gains, if these zeros end up close to the dominant poles (roots of the denominator) of the system, they can cause a considerable overshoot.

By making a change in the controller structure as shown in Figure 6.27, we can eliminate one of these zeros.

Now, the derivative gain acts on the actual system output directly as opposed to acting on the error signal as before. We can find the CLTF of the new structure by first replacing the velocity minor loop (Figure 6.27b) with its closed-loop transfer function CLTF_v given by

$$\text{CLTF}_v = \frac{G}{1 + GK_d s}$$

$$= \frac{K_m}{ms^2 + (b + K_m K_d)s} \tag{6.21}$$

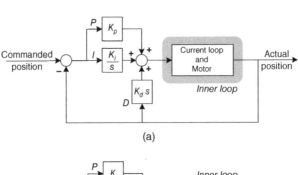

(a)

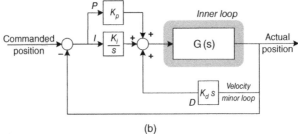

(b)

Figure 6.27 New PID control structure obtained by moving the derivative gain to the feedback loop. (a) Derivative gain on the feedback loop. (b) Same structure drawn in a different way

The CLTF$_v$ now corresponds to the G(s) in Figure 6.26. The controller transfer function in Figure 6.26 is now

$$C(s) = K_p + \frac{K_i}{s} \tag{6.22}$$

The CLTF for the entire system can be found as

$$CLTF = \frac{C(s)CLTF_v}{1 + C(s)CLTF_v} \tag{6.23}$$

After substituting Equations (6.21) and (6.22) into (6.23) and some algebra, we obtain

$$CLTF = \frac{K_m(K_p s + K_i)}{ms^3 + (b + K_m K_d)s^2 + K_m K_p s + K_m K_i} \tag{6.24}$$

If we compare Equations (6.20) and (6.24), we can see that their denominators are the same. Since the denominator of a transfer function governs the dynamic behavior of the system, the change in the control structure did not affect the dynamics of the system. However, the numerator polynomial is now only first order, hence, there is only one zero left in the numerator. As you may have already noticed, the numerator polynomial in Equation (6.24) and the $C(s)$ transfer function in Equation (6.22) represent a PI controller. Our new PID structure is nothing but the cascaded velocity/position loops shown previously in Figure 6.12.

As mentioned earlier, high gains can cause instability, especially the integral gain in a PID controller. A popular solution in industrial motion controllers is to use the integrator selectively only when in position as shown in detail later in Example 6.5.2. This means that the integrator is reset and turned OFF during the motion. When the motion is completed and the system stops, the integrator is turned back on to eliminate any remaining small errors. In our new structure shown in Figure 6.27b, if we turn the integrator OFF during the motion, we will have the following CLTF

$$CLTF = \frac{K_m K_p}{ms^2 + (b + K_m K_d)s + K_m K_p} \tag{6.25}$$

This is equivalent to the unity feedback system, as shown in Figure 6.28, where the tranfer function is for a *Type-1* system since it has one integrator (the single *s* in the denominator multiplying the terms in the brackets. It corresponds to an integrator, which is $1/s$ in the Laplace domain).

Type-1 systems will have the following error when tracking constant velocity input [13]. Figure 6.29 shows a typical point-to-point move with a trapezoidal velocity profile. If the cascaded velocity/position loops are used with the integration-when-in-position option, the system will have the following error during the constant velocity operation of the trapezoidal profile.

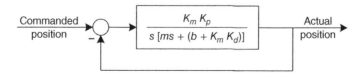

Figure 6.28   Type-1 system with unity feedback

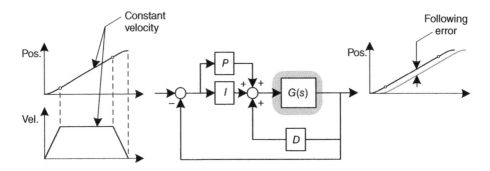

Figure 6.29   Constant velocity following error in a Type-1 control system

If we use the integrator during the motion, the following error may be eliminated but this time there will be the overshoot challenge. The proportional gain can be reduced to address the increased overshoot, but this will reduce the stiffness of the axis against disturbances.

6.2.3.2   Velocity Feedforward

If the input velocity profile is fed forward to the velocity loop directly, the system will become much more agile in responding to sudden changes in the desired trajectory. Any sudden change in the profile will directly be communicated to the velocity loop immediately by the feedforward gain allowing the system to react right away.

In an ideal system, the torque $T(t)$ should be equal to zero when traveling at constant speed. In real systems, the torque will be small and just enough to overcome the friction. Refering to Figure 6.30, to have $T(t) = 0$, the error $e_v(t)$ in the velocity loop must be zero.

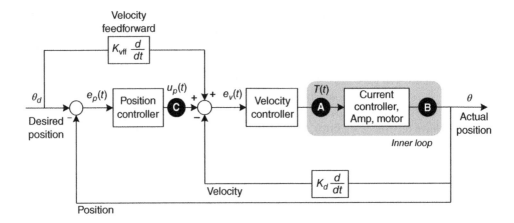

Figure 6.30   Velocity feedforward added to the cascaded velocity/position control structure

If a velocity feedforward with K_{vff} gain is used, as shown in Figure 6.30, we can write the following equation

$$e_v(t) = u_p(t) + K_{\text{vff}}\frac{d\theta_d}{dt} - K_d\frac{d\theta}{dt}$$

where $u_p(t)$ is the position controller output, which is a scaled version of the following error $e_p(t)$. Since we want $e_v(t) = 0$, if we set $K_{\text{vff}} = K_d$, then the last two terms will cancel out ($\theta_d \approx \theta$) and the following error, $e_p(t)$, will have to be equal to zero since $u_p(t)$ becomes zero. Hence, the velocity feedforward can eliminate the following error. However, if the K_{vff} gain is set too high, the velocity feedforward will cause overshoot because it will create large error spikes in the acceleration/deceleration regions of the velocity profile.

6.2.3.3 Acceleration Feedforward

The overshoot due to high-velocity feedforward gain can be addressed by adding acceleration feedforward. It eliminates the overshoot leading to an overall system with fast response and stiff disturbance rejection.

During the acceleration/deceleration, we need torque. The only way to get torque is to have $e_v(t) \neq 0$. But this is not desireable since it results in following error. Instead, the needed torque command comes into the velocity loop from the acceleration feedforward as shown in Figure 6.31. It is made directly proportional to the acceleration commanded by the input motion profile ($= K_{\text{aff}}\frac{d^2\theta}{dt^2}$) by adjusting the K_{aff} gain. If the K_{aff} gain is selected as

$$K_{\text{aff}} = \frac{m}{K_m}$$

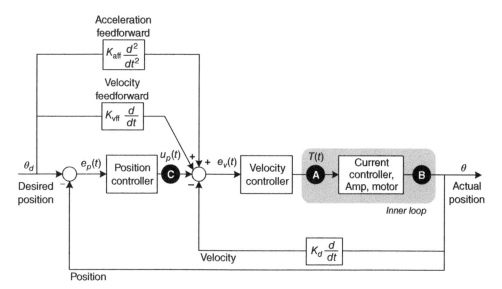

Figure 6.31 Acceleration and velocity feedforward added to the cascaded velocity/position control structure

then, we can write the following equation of motion

$$\frac{m}{K_m}\frac{d^2\theta_d}{dt^2}K_m = m\frac{d^2\theta}{dt^2}$$

where the left-hand side of the equation is the torque generated by the controller and the right-hand side is the resulting acceleration. From this equation, it is obvious that $\theta_d = \theta$, which means no following error. It should be noted that the K_{aff} gain is implicitly proportional to the inertia (mass). Therefore, it would need to be adjusted everytime the load inertia changed.

Not all industrial motion controllers have the acceleration feedforward feature. If this is the case, the overshoot from the high-velocity feedforward gain can be dealt with by reducing the other loop gains. The designer will have to trade-off between fast response and stiffness of the system.

6.2.3.4 Practical Implementation with Feedforward Gains

A common version of the cascaded velocity/position controller with the feedforward gains is shown in Figure 6.32. The Position Controller block in Figure 6.31 is replaced by a straight pass-through of the position error and its integral on a parallel branch. This is basically a PI controller with a slightly different implementation given by the following transfer function:

$$G_{PI}(s) = K_p\left(1 + \frac{K_i}{s}\right)$$

Note that the proportional gain K_p multiplies both terms and now appears in place of the Velocity Controller in Figure 6.31. You should compare this version to the "P"

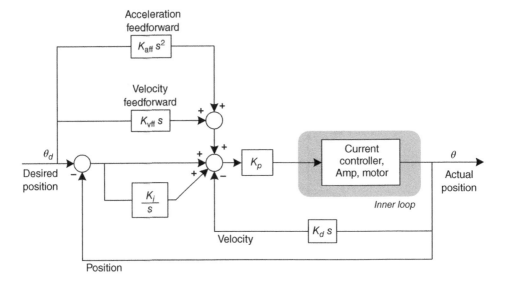

Figure 6.32   Cascaded velocity/position controller with velocity and acceleration feedforward gains

and "I" signal configurations in Figure 6.24 and the resulting controller transfer function in Equation (6.22) to see the slight differences. In fact, in Figure 6.32, the K_p gain actually multiplies all three control signals as

$$G_{\text{PID}}(s) = K_p \left(1 + \frac{K_i}{s} - K_d s \right)$$

The velocity feedforward branch differentiates the desired position profile and scales it by the K_{vff} gain to create its signal proportional to the commanded velocity. Similarly, the acceleration feedforward branch differentiates the desired position profile twice and scales it by the K_{aff} gain to create its signal proportional to the commanded acceleration.

6.3   Inner Loop

Motion control servo systems utilize an inner loop and one or more outer loops to achieve the desired control performance. These are feedback loops used to create a closed-loop control system for each axis. Figure 6.12 shows a generic implementation of such a control system for one axis.

The inner loop is also called the *current loop*. This feedback loop regulates the torque by controlling the currents to the motor. As shown in Figure 6.33, desired current is the input for the inner loop. The difference of the desired current from the actual current fed back from the motor is the instantaneous current error. This error is the input to a current controller which then generates the necessary voltage commands for the legs of the inverter. These voltages are usually in the form of PWM signals as explained in Section 6.1.2.2. The inverter applies the phase voltages to the motor. Phase currents are measured and fed back.

The design of the outer loops (velocity and position loops in Figure 6.12) is somewhat general and independent of the type of motor being controlled. However, the inner loop is specific to the type of motor being controlled [16]. In the following sections, we will first look at the inner loops used to control the two types of motors presented in this book. The inner loop design specific to a motor type replaces the generic inner loop in Figure 6.12 between points "A" and "B."

Torque produced by a DC *brushed* motor is proportional to the current into the terminals of the motor. Since this type of motor commutates itself, it is fairly easy to control the torque by just adjusting the amount of current supplied to the motor. In the case of brushless motors, such as the brushless DC, AC servo, or AC induction motors, the currents and voltages applied to the motor windings must be controlled accurately and independently as a function of the rotor position to produce torque.

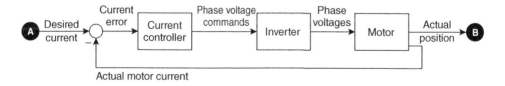

Figure 6.33   Generic structure of the inner loop (current loop)

In the brushless motors, the stator produces a rotating magnetic field. Torque is generated from the attraction or repulsion between the stator magnetic field and the rotor field. Maximum torque is produced when these two fields are kept perpendicular to each other during the operation of the brushless motors.

The stator field can be modeled in terms of phase currents. A *current space vector* is defined as a vector whose magnitude is proportional to the amount of current in a phase winding and whose direction is the same as the direction of the magnetic field generated by the winding. The current space vector allows us to show the combined effect of the phase currents as a single resultant vector. Figure 6.34 shows a three-phase AC servo motor. The current in each phase winding is shown as a current space vector. The direction of each vector is the same as the direction of the field generated by that winding. The resultant stator current space vector can be found by vector summation and is aligned in the same direction as the stator field.

We can decompose the resultant stator current space vector into its direct and quadrature components along the *dq*-axes. The rotor field is aligned with the *d*-axis. The quadrature current component produces torque since it will be perpendicular to the rotor field vector. The direct current component wastes energy and presses the rotor radially into its bearings. Therefore, the drive controller must function to minimize the direct component and maximize the quadrature component of the stator field to produce the most possible torque. The inner loop continuously regulates the individual phase currents to achieve this desired stator field alignment as the rotor is spinning.

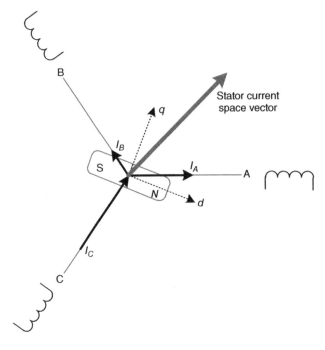

Figure 6.34   Stator current space vector

6.3.1 Inner Loop for AC Induction Motors

In an AC induction motor with a squirrel cage, three-phase voltages are supplied to the stator windings. The resulting stator field induces currents in the rotor bars which also generate the rotor field. The rotor magnetic field tries to keep up with the rotating stator field. Their inter-action generates torque. In a DC brushed motor, the torque can be controlled by adjusting the current into the motor terminals. In the AC induction motor such direct control of the rotor currents from an external source is not possible since they are induced by the interactions of the two fields.

Vector control or *field-oriented control* aims to make controlling an AC induction motor similar to a DC motor. This is achieved by controlling the current space vector directly in the *dq*-frame of the rotor. As explained in Section 6.3, the current space vector can be decomposed into direct and quadrature components. The quadrature component is the torque generating current. The direct component, also called the *magnetization current*, determines the torque gain K_T of the motor [4]

$$T = K_T(i_{ds}) i_{qs}$$

This equation is just like a DC motor torque equation ($T = K_T i$). In general, a constant torque gain is desired. Therefore, the magnetizing current needs to be kept constant.

Two PI controllers are used to control the quadrature and direct current components inde-pendently as shown in Figure 6.35. Here, i^*_{qs} corresponds to the desired current (or torque) in Figure 6.33. The desired direct current component is i^*_{ds}. As long as the direct current com-ponent maintains alignment with the rotor flux field and the quadrature current component is kept perpendicular to the direct component, the flux and torque can be controlled separately and the AC motor will produce torque-speed characteristics that are similar to a DC brushless motor.

The vector control algorithm manipulates the phase voltages and motor currents in the *dq*-frame. This requires the mathematical transformation of these quantities into the *dq*-frame before closed-loop control is applied by the PI controllers. Similarly, after the necessary con-trol signals are computed, they have to be transformed back to the three-phase stator voltages to be applied by the PWM inverter to the motor. These mathematical calculations involve Park

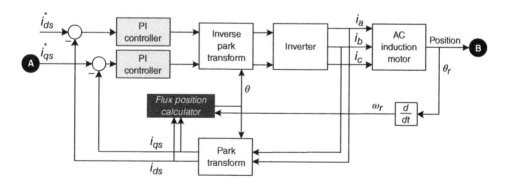

Figure 6.35   Inner loop implementation for AC induction motor using vector control

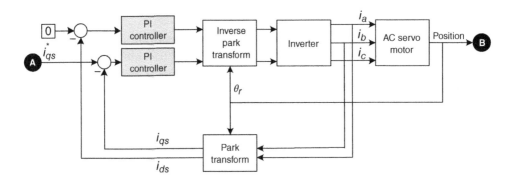

Figure 6.36   Inner loop implementation for AC servo motor using vector control

transform and its inverse as discussed in Section 4.5.1 and require fast digital signal processors (DSP) in the motion controller.

The alignment of the direct current component with the rotor flux field is challenging. The induced rotor magnetic field in an AC induction motor is not fixed to the rotor. There is a slip between the rotor and its field. Therefore, measuring the rotor position with an encoder is not sufficient to determine the angle between the stator field and the rotor field since the encoder would be measuring only the position of the rotor itself. If the motor is equipped with flux sensors mounted in the airgap, then the rotor flux vector direction can be measured. This approach is called *direct field-oriented control*. In this book, we explore the more common *indirect field-oriented control*, where the rotor field vector orientation is calculated (Figure 6.35). The details of these calculations are explained in Section 6.4.1.1.

6.3.2   Inner Loop for AC Servo Motors

The vector control method, as shown in Figure 6.35, can be used to easily control an AC servo motor. In an AC servo motor, the rotor has permanent magnets that produce the flux. Therefore, to control an AC servo motor, the desired direct current component (flux producing current) is set to zero. This regulates the d-axis current component to keep it at zero and, hence, forces the current space vector to be exclusively in the quadrature direction. The q-axis current component is the only current that produces useful torque; hence, the torque efficiency of the motor is maximized. In AC servo motors, there is no slip since the rotor flux is fixed to the rotor and rotates with it. As a result, there is no need for the flux position calculator. Figure 6.36 shows the vector control method with these changes.

6.4   Simulation Models of Controllers

Simulation models for each of the drive control structures were built using Simulink® . In each case, the model consists of a velocity outer loop and an inner loop specific to the type of motor being controlled. The velocity outer loop implements the block diagram in Figure 6.13. As discussed in Section 6.2, a second outer loop for position control can also be added to these models.

6.4.1 Simulation Model for Vector Control of an AC Induction Motor

The velocity loop implementation is shown in Figure 6.37a. In this case, the input to the model is a trapezoidal desired velocity in rad/s. The PI Controller block implements a PI velocity controller. The input to the PI controller is the velocity error. The output of the block is the torque demand (desired current command) to the inner loop. The output of the Inner loop and AC induction motor block is the actual motor shaft position which is differentiated and fed back. The position is also displayed in a scope.

The inner loop details and the AC induction motor are shown in Figure 6.37b. The inner loop implements the vector control algorithm, as shown in Figure 6.35. The encoder mounted on the motor shaft measures the rotor position. However, this is not sufficient to determine the angle between the stator and rotor field since the rotor field slips. The rotor field, or the *rotor flux direction*, needs to be calculated so that the vector control algorithm can align the direct current component with the rotor field and keep the quadrature component perpendicular to the direct component to enable DC motor-like torque control.

In the next section, we will look at how the rotor flux direction can be calculated. This will be followed by the details of the simulation model.

6.4.1.1 Vector Control Algorithm

Equations governing the dynamic behavior of an induction motor were given in Section 4.5.2. We can write the rotor voltage equation given at the bottom of Equation (4.28) explicitly as [4, 12, 15]

$$V_{dr} = R_r\, i_{dr} + \frac{d\lambda_{dr}}{dt} - (\omega_k - \omega_{re})\, \lambda_{qr} \tag{6.26}$$

$$V_{qr} = R_r\, i_{qr} + \frac{d\lambda_{qr}}{dt} + (\omega_k - \omega_{re})\, \lambda_{dr} \tag{6.27}$$

Also, the rotor flux linkages given in Equation (4.29) can be written explicitly as

$$\lambda_{dr} = L_m\, i_{ds} + L_r\, i_{dr} \tag{6.28}$$

$$\lambda_{qr} = L_m\, i_{qs} + L_r\, i_{qr} \tag{6.29}$$

If the *dq*-frame is fixed to the magnetic field of the rotor, then $\lambda_{qr} = 0$, and from Equation (6.29)

$$i_{qr} = -\frac{L_m}{L_r}\, i_{qs} \tag{6.30}$$

Since $\lambda_{qr} = 0$, its derivative will also be zero. In addition, both V_{dr} and V_{qr} are set to zero since the model is simulating a squirrel cage induction motor where no external voltage is applied to the rotor windings since the rotor contains bars that are shorted by end caps. Substituting (6.30) into (6.27), we can solve for

$$\omega_k - \omega_{re} = \frac{L_m}{\lambda_{dr}} \left(\frac{R_r}{L_r} \right) i_{qs} \tag{6.31}$$

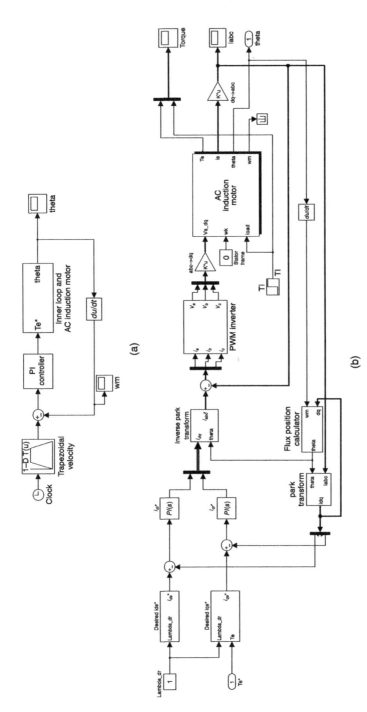

Figure 6.37 Simulation model for the velocity loop and inner loop for an AC induction motor. (a) Velocity loop for AC induction motor. (b) Inner loop details for vector control of an AC induction motor

Note that $\omega_s = (\omega_k - \omega_{re})$ is the *slip frequency*. Also, $\tau_r = \left(\frac{L_r}{R_r}\right)$ is the *electrical time constant of the rotor*. From Equation (6.26)

$$i_{dr} = -\left(\frac{1}{R_r}\right)\frac{d\lambda_{dr}}{dt} \tag{6.32}$$

If we substitute Equation (6.32) into (6.28), we obtain

$$\tau_r\frac{d\lambda_{dr}}{dt} + \lambda_{dr} = L_m i_{ds}$$

If i_{ds} is kept constant, then this equation can be simplified into

$$\lambda_{dr} = L_m i_{ds} \tag{6.33}$$

Substituting this equation back into (6.31) yields

$$\omega_s = \left(\frac{R_r}{L_r}\right)\frac{i_{qs}}{i_{ds}} \tag{6.34}$$

The slip angle θ_s can be found by integrating Equation (6.34). If the mechanical position of the rotor, θ_r, is measured relative to the stator (usually using an encoder on the motor), then we can add the two angles to find the orientation of the rotor field

$$\theta = \theta_s + \theta_r$$

Here, θ is the angle between the d-axis of the dq-frame and the stator frame. This angle is used in current commutation to maintain the desired vector orientations between the stator and rotor frames.

The motor torque is given by [14]

$$T_e = \frac{3}{2}\left(\frac{P}{2}\right)\left(\frac{L_m}{L_r}\right)(\vec{\lambda}_r \otimes \vec{i}_s)$$

$$= \frac{3}{2}\left(\frac{P}{2}\right)\left(\frac{L_m}{L_r}\right)([\bar{R}]\lambda_r \odot \vec{i}_s)$$

$$= \frac{3}{2}\left(\frac{P}{2}\right)\left(\frac{L_m}{L_r}\right)(\lambda_{dr}i_{qs} - \lambda_{qr}i_{ds})$$

Since $\lambda_{qr} = 0$, it simplifies as

$$T_e = \frac{3}{2}\left(\frac{P}{2}\right)\left(\frac{L_m}{L_r}\right)\lambda_{dr}i_{qs} \tag{6.35}$$

This equation can be rewritten as

$$T_e = K i_{qs} \tag{6.36}$$

where $K = \frac{3}{2}\left(\frac{P}{2}\right)\left(\frac{L_m}{L_r}\right)\lambda_{dr}$. If λ_{dr} can be kept constant, then K becomes constant. Then, Equation (6.36) looks like the torque equation of a DC motor where torque is directly proportional to the current.

Given required rotor flux λ_{dr} and torque demand T_e, the indirect field-oriented vector control algorithm creates current commands for the motor in the dq-frame from Equations (6.33) and (6.35) as

$$i_{ds} = \frac{1}{L_m}\lambda_{dr} \tag{6.37}$$

$$i_{qs} = \frac{2}{3}\left(\frac{2}{p}\right)\left(\frac{L_r}{L_m}\right)\left(\frac{T_e}{\lambda_{dr}}\right) \tag{6.38}$$

The position of the rotor is measured and added to the integral of the slip frequency in Equation (6.34) as

$$\theta = \int \left(\frac{R_r}{L_r}\right)\frac{i_{qs}}{i_{ds}}\,dt + \theta_r \tag{6.39}$$

to estimate the rotor field orientation. This, in turn, is used in coordinate transformations using the Park and inverse Park transforms to compute the required stator phase voltages (V_{abc}).

6.4.1.2 Simulation of the Inner Loop for Vector Control

The vector control algorithm manipulates the currents and phase voltages in the dq-frame. As a result, the PI current controllers work on DC signals. This leads to equally good performance at low and high speeds.

The inner loop for vector control, as shown in Figure 6.35, requires two input commands: (1) desired d-axis current component (i_{ds}^*) and (2) the desired current (or torque) denoted by i_{qs}^*. The `desired ids*` block and the `desired iqs*` block on the left-hand side of Figure 6.37b compute these inputs to the inner loop using Equations (6.37) and (6.38). Here, the required rotor flux, λ_{dr}, is set at a certain value to define the required magnetization current. The torque demand, T_e, comes from the velocity controller on the outer loop. Details of the `desired ids*` and `desired iqs*` blocks are shown in Figure 6.38.

The PI current controllers in the `PI(s)` blocks act on the current error in the dq-frame, which is the difference between the desired currents and the actual currents. These errors are calculated using actual currents measured from the motor and transformed into the dq-frame. The outputs of the PI current controllers are transformed into the abc-frame using the inverse

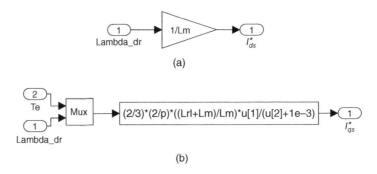

Figure 6.38 Details of the desired ids* and desired iqs* current blocks. (a) Desired d-axis current component, i_{ds}^*. (b) Desired q-axis current component, i_{qs}^*

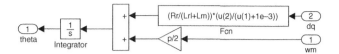

Figure 6.39   Details of the flux position calculator block

Park transform. At this point, just on the right-hand side of the `Inverse Park Trans-form` block, we have three-phase currents as references for the PWM inverter. The `PWM Inverter` block generates phase voltages (`Va`, `Vb`, `Vc`) for the motor based on the current error in each phase. In this model, we used the same AC induction motor model that was developed in Section 4.5.2. This model was in the *dq*-frame. Therefore, three-phase voltage output of the PWM inverter had to be converted back into the *dq*-frame using the `abc->dq` block to be used with the `AC induction motor` block. The resulting rotor mechanical speed is used in the `Flux Position Calculator` block.

The `Flux Position Calculator` block estimates the rotor field orientation using Equation (6.39) as shown in Figure 6.39. The flux position is then fed to the `Inverse Park Transform` block and used in the calculation of the three-phase current commands. In both Figures 6.38b and 6.39, a small nonzero quantity (1×10^{-3}) was added to the denominator of the equations to avoid a divide-by-zero error at the beginning of the simulation.

6.4.2   Simulation Model for Vector Control of an AC Servo Motor

The velocity outer loop, as shown in Figure 6.40a, is the same as the induction motor control model. The inner loop details and the AC servo motor are shown in Figure 6.40b. The inner loop implements the modified vector control algorithm as shown in Figure 6.36. One difference of this control structure from the vector control of an AC induction motor is that there is no flux position calculator since the rotor flux is fixed to the rotor. The second difference is that the desired direct current component is set to zero since permanent magnets produce the flux.

As shown in Figure 6.40b, two PI current controllers in the `PI(s)` blocks regulate the currents in the *dq*-frame. Then, the inverse Park transform is used to convert the currents into three-phase currents in the abc frame. The `PWM inverter` block generates phase voltages (`Va`, `Vb`, `Vc`) for the motor. In this model, we used the same AC servo motor model that was developed in Section 4.5.1. At the output of the `AC Servo Motor` block, the motor phase currents are measured, converted into *dq*-frame currents using the Park transform and fed back to the PI controllers.

6.5   Tuning

Tuning is the process of adjusting controller gains so that the servo system can follow the input command as closely as possible. The performance of the servo system can be quantified in terms of how quickly and accurately it can follow the input and how stable the response is.

A servo system can be analyzed in *time domain* or in *frequency domain*. The frequency domain analysis relies on *Bode plots*, whereas the time domain analysis employs *step response*.

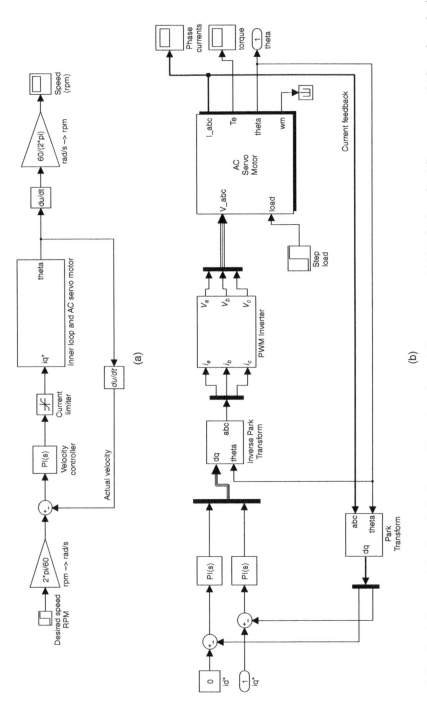

Figure 6.40 Simulation model for the velocity loop and inner loop for an AC servo motor. (a) Velocity loop for AC servo motor. (b) Inner loop details for vector control of an AC servo motor

When tuning the controller of a machine, generally it is challenging to obtain a Bode plot unless instruments, such as a DSA, are available to measure the frequency response of the machine and generate the Bode plots. Some motion controllers may have built-in frequency domain analysis capability, but it is not a common feature. However, all controllers come with the capability of generating a step input and capturing the resulting response.

The command response can be measured using the *settling time*. It shows us how quickly the system can follow an input. A step input is given to the system and the resulting response is captured. Settling time is defined as the time it takes for the system response to enter *and stay* within ±2% of the command [13]. Figure 6.41 shows a system where the command is 100 rpm and the settling time is 0.096 s. Notice that the response enters the 2% envelop near the 100 rpm target first at about 0.06 s but does not stay in the envelop. The first time it enters and stays in the envelop afterwards is at 0.096 s. Slower or more sluggish systems will have longer settling times.

In practice, motion control systems would not need to respond to step input commands. For example, the input commands generated for the velocity loop by the position outer loop are much gentler commands such as a trapezoidal velocity profile. In a way, the step response represents the worst-case scenario a system may encounter. If the system can be tuned to perform at a satisfactory level for a step input, it will most likely do even better under the normal, gentler operating conditions.

A servo system can be considered stable if its response to a command is predictable. Excessive overshoot or sustained oscillations are considered to be unstable system behaviors. In the time domain, *percent overshoot* is often used as a measure to quantify stability. Percent overshoot (OS%) is defined as the ratio of difference of the peak response from the commanded input to the commanded input. In Figure 6.42a, the system has 13% overshoot ($= (113 - 100)/100$). The commanded input was 100 rpm. The difference of the peak response from the commanded input was 13 rpm. Similarly, the system in Figure 6.42b has 36% overshoot. Even though both systems are stable, the one shown in Figure 6.42b with the 36% overshoot has a lesser *margin of stability*. In other words, it is closer to becoming unstable, if operating conditions change.

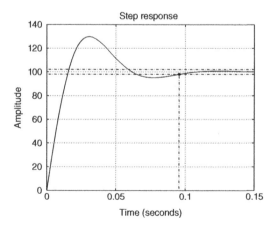

Figure 6.41   Settling time as a measure of responsiveness

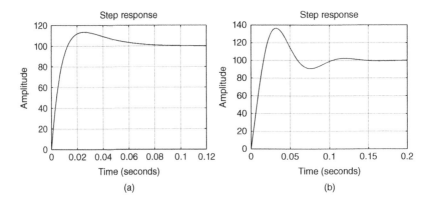

Figure 6.42 Percent overshoot as a measure of stability. (a) System response with 13% overshoot
(b) System response with 36% overshoot

A servo system tries to reduce or eliminate the error between the commanded input and the
resulting system response as it tries to follow the input. The *gain* of the control system determines how hard the system tries to reduce the error. If the gain is high, the system generates
large torques for even the smallest error to try to correct the error quickly. A motion control
system accelerates and decelerates a load during motion. As the system tries to follow a change
in the input command, the inertia and the high gains will lead to overcorrection. As a result,
the system may oscillate around the desired target position. These oscillations are also known
as *ringing* and are considered an instability.

Each application requires a balance between stability and responsiveness. High-gain settings
not only improve responsiveness but also make the system less stable by increasing overshoot.
For a more stable system, the gains should be kept low. However, this time the resulting system
will have a sluggish response and will not be stiff when subjected to an external disturbance
such as a sudden change in the load. Tuning is all about striking a balance between these
competing requirements to obtain an optimal performance in a specific application. Generally
speaking, *a well-tuned servo system is stiff and has the fastest possible response with little or
no overshoot and steady-state error.*

Saturation of the control signal must be avoided during tuning. To tune a system, step or
trapezoidal input commands are used. The amplitude of the input command should be selected
to avoid the saturation of the control signal (current command input to the amplifier). If the
input command is repeating, such as a square wave for step input, then its frequency should be
fairly low so that the system can follow it when properly tuned. Typically, the current command
(control signal) sent to the power amplifier is a ± 10 V signal where $+10$ V is the full positive
torque (or current) and the -10 V is the full negative torque command to the motor. If the
tuning of the gains results in approaching or exceeding these values during motion, especially
for extended periods of time, then this is an indication of the system being pushed too hard.
To avoid saturation, the amplitude of the command input should be reduced. For example, in
the case of a move with trapezoidal velocity, the operational speed at the flat section of the
trapezoid can be reduced, along with the acceleration and deceleration rates so that the system

is not pushed too hard during the move. Most motion controllers have software features to monitor the control signal while tuning.

6.5.1   Tuning a PI Controller

The performance requirements, such as the settling time and allowable overshoot, vary from application to application. Therefore, system designers will need to make the necessary adjustments based on the hardware capabilities and these requirements. However, as a general approach, the following steps can be used to tune a PI controller [8].

PI TUNING METHOD

---

1. Set $K_i = 0$ and K_p low,
2. Apply square wave as input command,
3. Increase K_p slowly until the response reaches the square wave input target with little or no overshoot,
4. Increase K_i for up to 15% overshoot, and
5. Verify that the current command signal does not saturate during executing motion for the application. If saturation occurs, especially for extended periods of time, the system is pushed too hard. Turn down the gains until the saturation can be avoided.

In this procedure, the square wave represents the theoretical step input. The rising edge of the square wave is just like the theoretical step input. The falling edge is also like the step input except it is a command in the opposite direction. The amplitude of the input square wave should be selected to avoid saturating the control signals. Also, the frequency of the square wave should not be too high so that the system will have sufficient time to respond to the input. As shown in Section 6.2.1.1, Equation (6.7), the two controller gains interact with each other. Hence, adjusting one will affect the other and the overall system response. Therefore, after following the procedure above, it may still be necessary to make additional "fine tuning" adjustments to the gains.

In the cascaded velocity–position loop control structure, each outer loop contains a PI controller. The tuning procedure outlined above can be applied to each of these loops sequencially starting from the velocity loop. Each loop must be tuned to its maximum performance capability since its performance will impact the performance of the next loop to be tuned. Typically, the bandwidth of each loop is 20–40% of the previous loop it encloses [8]. The inner (current) loop must have the fastest response (largest bandwidth). The same procedure can also be used in tuning the PI controller of the current loop. However, it may not always be possible to obtain the 10–15% overshoot in the last step of the procedure depending on the design of the power electronics in the drive. Still, the K_p gain increases the responsiveness of the current controller, while the K_i gain provides stiffness to the current loop to reject any disturbances.

As mentioned before, step input represents the worst-case scenario. Normally, real systems would not be subjected to such inputs during regular operation. Typically, after tuning the system with the step input, the system performance is tested using more realistic inputs such as trapezoidal or S-curve velocity profiles.

Sustained following error can generate a large integral signal due to the error accumulation in the integrator over time. Two methods to keep the integral signal at a reasonable level are discussed in Section 6.5.2.1.

■ **EXAMPLE 6.5.1**

An axis with an AC servo motor is controlled using the velocity loop given in Figure 6.40a. The total inertia, including the reflected load inertia, is $J = 8 \times 10^{-4}\,kg - m^2$. The motor has eight poles and $R = 2.9\,\Omega, L = 11\,mH$. Tune the PI velocity controller so that the system response can follow a velocity step input as fast as possible with 15% or less overshoot.

Solution
We need to start from tuning the inner (current) loop first. To do this, a square wave needs to be supplied to the inner loop as a current step input command. The resulting current response must be captured. In commercial motion controllers, there is a way to "open" or disable the velocity loop and provide the square wave input to the current loop. To achieve the same thing with the simulation model explained in Section 6.4.2, the velocity loop was modified as shown in Figure 6.43. A current input switch was used to select the input signal for the inner loop. Another switch called velocity input switch was also added to test the performance of the system with trapezoidal velocity input after the tuning has been completed.

To tune the current loop, the current input switch is put in the up position. This opens (disables) the velocity loop. The input for the current loop comes from the square wave block. The square wave current command was set to 1 A with 25 Hz frequency. This signal is connected to the i_q^* quadrature current input to the inner loop in Figure 6.40b. The resulting current response was also measured (not shown) from the i_q current feedback signal in the inner loop.

Our goal is to tune the K_p and K_i gains of the PI (s) controller for the i_q current loop, as shown in Figure 6.40b. As an initial setting, $K_p = 3, K_i = 0$ was tried. As seen in Figure 6.44a, even though the current response could follow the input it was oscillatory and not accurate. After additional experimentation with other gain settings, $K_p = 100, K_i = 500$ provided satisfactory current response as shown in Figure 6.44b. The current loop can now follow the input square wave very accurately and the response is very fast providing a large bandwidth.

To tune the velocity loop, the current input switch is put in the down position. This closes the velocity loop and disconnects the square wave input to the current loop. The velocity input switch is switched to the desired speed RPM block. This is a step input set to 200 rpm. In commercial motion controllers, this input is often a square wave, too. In this step of the tuning process, our goal is to adjust the K_p and K_i gains of the PI (s) velocity controller in the velocity loop.

Following the procedure outlined in Section 6.5.1, K_i was set to zero and the proportional gain was set to a low value ($K_p = 0.03$). As shown in Figure 6.45a, this resulted in a slow and sluggish speed response. Then, the K_p gain was slowly incremented and the system response

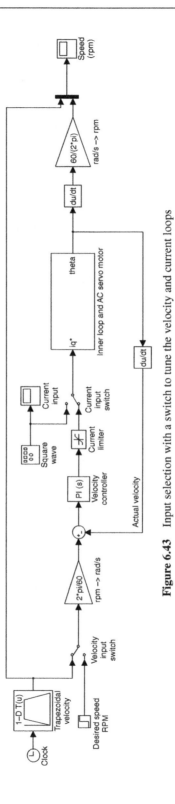

Figure 6.43 Input selection with a switch to tune the velocity and current loops

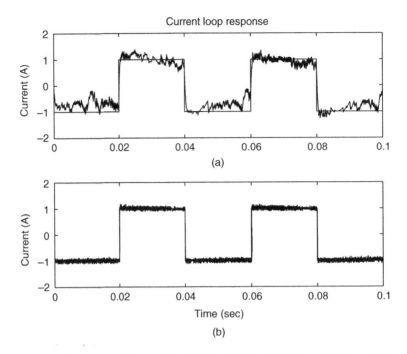

Figure 6.44   Current loop tuning. (a) $K_p = 3, K_i = 0$; (b) $K_p = 100, K_i = 500$

was monitored. At $K_p = 2.5$, the fastest possible response without any overshoot was obtained as shown in Figure 6.45b. Any slight increment beyond this setting (e.g., $K_p = 2.7$) resulted in overshoot. Finally, while keeping $K_p = 2.5$, the K_i gain was incremented and the response was monitored in each new setting. At $K_p = 2.5, K_i = 300$, the tuning process was finalized where the system response had 15% overshoot as shown in Figure 6.45c.

How well will the system perform in normal operations? Typically, after tuning the velocity loop using the step input, the designer will test the system with more realistic inputs, such as a trapezoidal velocity input. To test our simulated system, we use the `velocity input switch` (Figure 6.43). When this switch is put in the up position, a trapezoidal velocity profile is provided as the input velocity command. This profile accelerates the system to the desired 200 rpm operational speed, keeps it at this speed for a while and decelerates it to a stop.

To show the impact of the tuning on the system performance, the system response was simulated first using the initial gain settings and then with the tuned gains. Figure 6.46a shows the system performance with $K_p = 0.03, K_i = 0$. As can be seen, the system is not able to track the desired velocity at all. Figure 6.46b shows the same system with the tuned gains of $K_p = 2.5, K_i = 300$. The system can track the desired velocity very well. In fact, looking at the figure, it is hard to distinguish the input and response signals from each other except for the slight overshoot at 0.02 s.

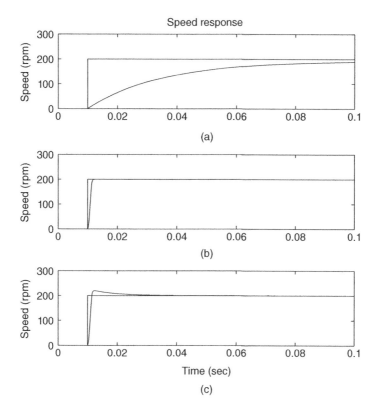

Figure 6.45 Velocity loop tuning. (a) $K_p = 0.03, K_i = 0$; (b) $K_p = 2.5, K_i = 0$; (c) $K_p = 2.5, K_i = 300$

6.5.2 Tuning a PID Position Controller

A PID controller is a PI controller with the addition of the derivative "D" control signal. The benefit of the additional signal is that it allows the proportional gain to be set higher than normal. The higher "P" gain will cause more overshoot, but the "D" gain will bring it down.

One of the challenges in the PID controller is the adverse effect of noise on the "D" gain. Noisy signals will create problems with the differentiation in the "D" signal of the PID controller. To solve this problem, a low-pass filter can be used in series in the derivative branch, but this can add unwanted lag in the controller. The best thing to do is to eliminate or reduce the source of the noise in the signals.

We will look at a basic method to tune a PID *position* controller. It should be noted that the method is only a procedure to follow to approach the tuning process systematically. It does not result in a unique set of gains. In other words, after following this method, the set of values found for the gains is not necessarily the only possible or optimal set of gains for the system in hand. Other combinations may also give satisfactory performance.

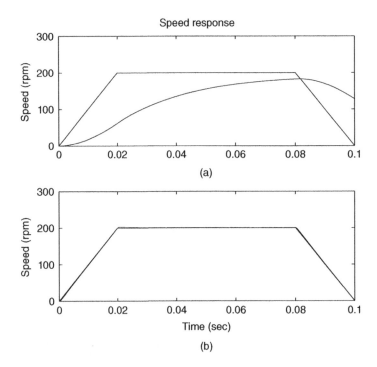

Figure 6.46 Tracking trapezoidal velocity input. (a) initial tuning with $K_p = 0.03, K_i = 0$; (b) tuned velocity controller with $K_p = 2.5, K_i = 300$

6.5.2.1 Integrator Saturation

Sustained following error can generate a large integral signal due to a high K_i gain and the error accumulation in the integrator over time. The main goal of the integrator is to generate enough current command to the amplifier to overcome the following error due to friction. Two common methods to keep the integrator signal at a reasonable level are:

1. Clamping the integrator output signal
2. Activating the integrator only when stopped (in position).

Overcoming friction requires a certain amount of additional push from the motor after it completes the motion. In the first approach, the integrator output is clamped (or limited) to a relatively low value just enough to overcome the friction. When the friction causes a position error, the error starts to accumulate in the integrator. Soon it reaches to the level of the clamp and the integrator can output just that level of current command. At this point, the motor overcomes the friction and starts moving without accumulating the integrator signal excessively.

In the second method, the integrator is deactivated during motion. This allows us to set much higher K_p and K_d gains and avoid overshoot or oscillations caused by the integrator while in motion. Once the commanded velocity is zero (stopped), the integrator is activated. At this point, if there is a position error due to friction, it will be a small amount which starts to accumulate in the integrator as time passes to create the additional push needed by the motor to overcome the friction and eliminate the error. If a new move is commanded, the integrator is reset and deactivated during the move.

Most controllers have features to allow these integrator anti wind-up settings. If implemented, these techniques allow setting the K_i gain much higher than otherwise possible resulting in a faster cancellation of the position error without the unwanted side effects of the integrator [7].

6.5.2.2 Basic PID Tuning Method

In this method, a square wave is used as the input command to the control system. First, the K_p and K_d gains are tuned until a critically damped response can be obtained. Then, the K_i gain is introduced to complete the process [8].

BASIC PID TUNING METHOD

---

1. Set $K_p = 0, K_i = 0$ and K_d low,
2. Apply square wave as an input command for the desired position. The square wave represents the theoretical step input.
3. Increase K_p slowly until the response reaches the square wave input target with about 10% overshoot and no oscillations.
4. Increase K_d to eliminate the overshoot as much as possible to approach or to obtain a critically damped response.
5. Increase K_i with small increments while maintaining the critically damped response. If overshoot is observed, keep it under 15%.

   Tuning the integral gain is difficult because it can cause overshoot even with small amounts of K_i. Integrator saturation will generate a large integral signal resulting in a big overshoot or oscillations. Therefore, integrator saturation must be avoided using clamping or activating the integrator *only when in position*. Most controllers have a feature to allow these types of integrator anti-wind-up settings [7].
6. Verify that the current command signal does not saturate during executing motion for the application. If saturation occurs, especially for extended periods of time, the system is pushed too hard. Turn down the gains until the saturation can be avoided.

■ **EXAMPLE 6.5.2**

The same axis in Example 6.5.1 is now being controlled by a PID position controller as shown in Figure 6.47a. The axis has a friction torque of $T_f = 0.1\,\text{Nm}$. Tune the position controller using the basic PID tuning method to obtain a fast response with little or no overshoot.

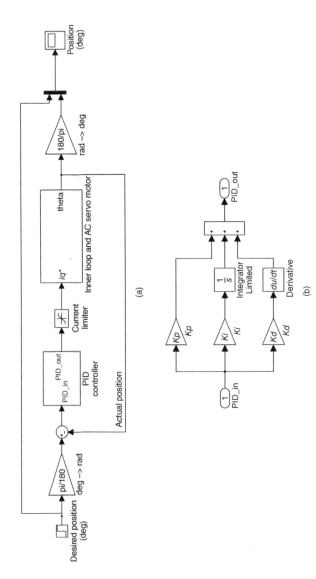

Figure 6.47 AC servo motor controlled by a PID position controller (a) Single loop PID position controller with AC servo motor. (b) PID controller details

Solution

It is assumed that the current controller has already been tuned as explained in Example 6.5.1. To start the position controller tuning process, first the proportional and integral gains are set to zero ($K_p = 0, K_i = 0$). The derivative gain is set to a low value of $K_d = 3$. A step input of $90°$ ($\frac{1}{4}$ turn) is applied to the system. Initially, since $K_p = 0$, no response is observed. Then, K_p is increased and the response is monitored until a response with 10% overshoot is obtained as shown in Figure 6.48a. Now K_d is increased incrementally until the overshoot is eliminated and a critically damped response is obtained as shown in Figure 6.48b.

The K_i gain can help eliminate the remaining little bit of error in the system which may bring the system to ± 1 count around the target. The "I" signal builds up by accumulating the following error over time while the system is in motion.

Continuous Integration

Usually a small amount of integrator signal is used since too much of it may lead to instability. In this example, the integrator gain was set at $K_i = 30$, which was as high as possible without causing any overshoot in the response as shown in Figure 6.48c. However, if the system is disturbed by a sudden change in the load as in Figure 6.49, it cannot recover quickly and eliminate the position error. The K_i gain must be increased for better recovery performance.

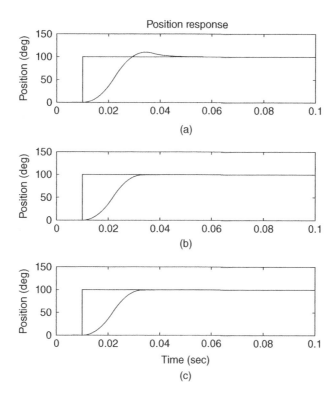

Figure 6.48 Tuning a PID position controller for an axis with an AC servo motor. (a) $K_p = 585, K_i = 0, K_d = 3$; (b) $K_p = 585, K_i = 0, K_d = 3.55$; (c) $K_p = 585, K_i = 30, K_d = 3.55$

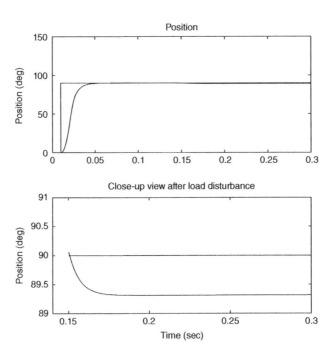

Figure 6.49 PID position controller for an AC servo axis with low K_i gain ($K_p = 585, K_d = 3.55, K_i = 30$). A load disturbance torque (8 Nm) at 0.15 s causes 0.7° following error, which cannot be eliminated by the controller

Figure 6.50 shows the system response with $K_i = 30$ and $K_i = 20\,000$. The following error is eliminated with $K_i = 20\,000$, but now the system has 30% overshoot, which is unacceptable.

Integrate When In Position
In this approach, the integrator is activated only when no motion is commanded. In other words, the integrator is activated only after the system settles near its target and therefore is almost in position. Since the error is small near the target, a relatively large value of K_i can be used without the usual side effect of overshoot and oscillations. The integrator will eliminate the small following error. As soon as motion is commanded again, the integrator is reset and disabled while in motion. Most motion controllers allow such a setting as a simple software feature. To simulate this feature, the integrator in the PID controller was set up to be triggered by an external trigger as shown in Figure 6.51b. The velocity was used as the trigger to activate the integrator when the velocity became zero.

Figure 6.52 shows the response with $K_i = 20\,000$. The disturbance occurs at $t = 0.15$ s. The system gets in position already by $t = 0.26$ s with the zero following error. It should be noted that the method allowed a high K_i setting making the system much more responsive to the disturbance, but the overshoot remained as it were with the original system, which was zero percent. The dynamic response (overshoot and settling time) was governed by the K_p and K_d gains, while the system was in motion. When it stopped, the K_i gain came into effect to overcome the following error. The velocity response had a 1.6% overshoot at the corners of the trapezoidal profile, but this is well below the acceptable 10% range in practice.

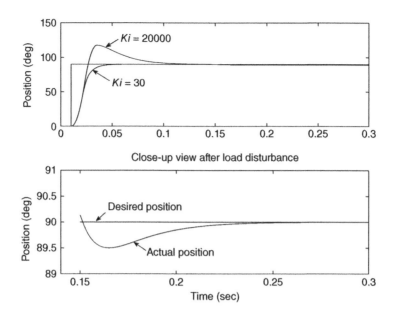

Figure 6.50 Tuning results with two different K_i settings using continuous integration. PID position controller for an AC servo axis with $K_p = 585, K_d = 3.55$. The following error due to the load disturbance is eliminated with $K_i = 20\,000$ but the overshoot is now 30%

The integrator gain could be set as high as $K_i = 50\,000$ still with no overshoot. The system eliminated the following error faster at around $t = 0.18$ s. Any further increase in K_i began to cause some slight oscillations about the target position.

It is very important to make sure that the current command to the drive is not saturated when the motor is used in its intended application. In this case, we used $\pm 10\,A$ as the saturation limits, which is typical for a drive for this size motor. As shown in Figure 6.53a, when the system is following a trapezoidal velocity profile, the phase currents do not saturate.

Figure 6.53b shows the torque generated by the motor during the trapezoidal move. It first steps up to accelerate the load. Then, during the constant speed travel, it falls down to 0.1 Nm to overcome the friction (hard to see in the figure). Next, it steps in the opposite direction to decelerate the load to a stop. At $t = 0.15$ s, suddenly the disturbance load is applied. The integrator rapidly increases the torque to 8 Nm so that the motor can continue to hold position while rejecting the 8 Nm load disturbance.

Integrator Clamping
In this approach, we set limits on the maximum amount of current the integrator can command. By altering these limits, the amount of contribution by the integrator gain can be adjusted. Often the main role of the integrator is to overcome the friction. Hence, the integrator output can be clamped at a level just sufficient to overcome the friction torque in the system and not much more. This enables us to reduce the potential overshoot and oscillation effects of the integrator.

Figure 6.54a shows the response with continuous integration and no clamping. As before, the integrator gain was set to $K_i = 20\,000$ to reject the disturbance but the overshoot is 30%.

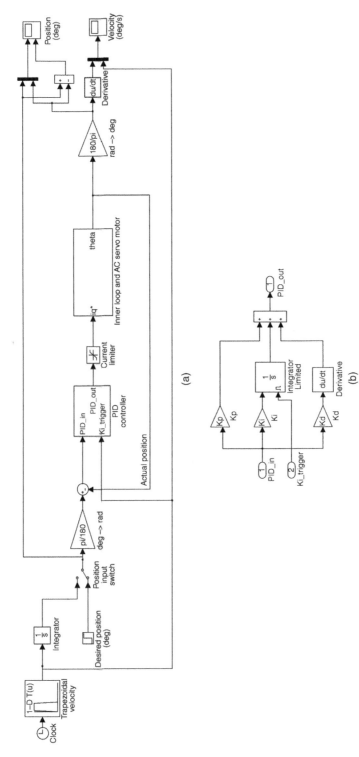

Figure 6.51 System setup for tuning the PID position controller for an AC servo axis using the integrate-when-in-position approach. (a) Control system configuration. (b) Inside the PID block. The integrator was set up to be triggered by an external signal (velocity)

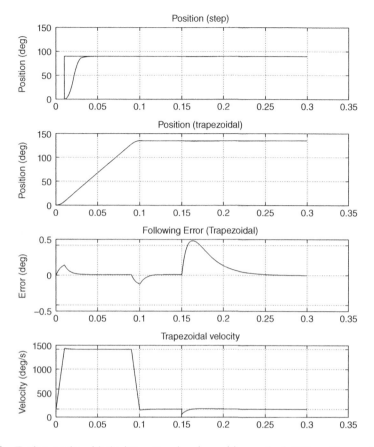

Figure 6.52 Tuning results with the integrate-when-in-position method. PID position controller for an AC servo axis with $K_p = 585, K_i = 20\,000, K_d = 3.55$

The same system can be improved significantly by clamping the integrator output at $\pm 8.5\,A$. This was done by entering ± 8.5 in the output limit parameters of the integrator in the PID controller block in the system. Note that in this simulation the integrator is continuously on. Therefore, the external trigger used in the previous method was removed. As seen in Figure 6.54b, the clamping allowed the gain to be increased up to $K_i = 50\,000$, which quickly rejects the disturbance and still provides a response with no overshoot as in the original low gain setting response in Figure 6.49.

Clamping + Integration When in Position
When these two approaches are combined, there can be an added safety benefit. Primarily, the integration-when-in-position approach is used to take out any remaining following errors. However, having clamps on the integrator output can provide added safety in case the mechanism is jammed or the axis hits a limit. This way, the continuing error accumulation cannot charge the integrator to dangerous levels passed the clamping limits.

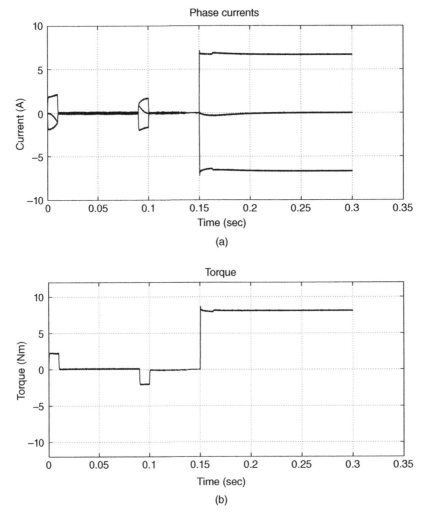

(a)

(b)

Figure 6.53 Phase currents and torque in trapezoidal move with the integrate-when-in-position method. PID position controller for an AC servo motor with $K_p = 585, K_i = 20\,000, K_d = 3.55$. (a) Phase currents. (b) Torque

6.5.3 Tuning a Cascaded Velocity/Position Controller with Feedforward Gains

As presented in Section 6.2.3.4, this control structure is based on a modified PID position controlller where the "D" signal was moved to the feedback loop. In addition, the controller has velocity and acceleration feedforward signals. Therefore, there are five controller gains that need to be tuned.

In this section, we will look at a procedure to tune this type of controller. It should be remembered that these procedures do not result in a unique set of gains. Instead, they describe a systematic approach to identify a set of gains that will satisfy typical performance

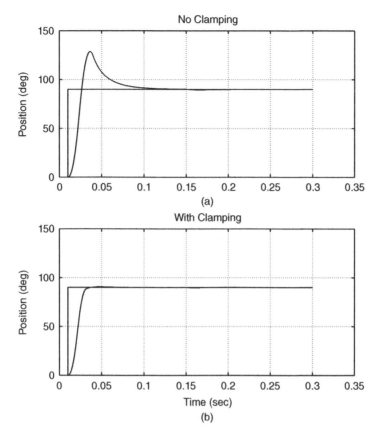

Figure 6.54 System response with continuous integration. (a) no integrator clamping and $K_p = 585, K_i = 20\,000, K_d = 3.55$; (b) integrator clamping with $\pm 8.5\,A$ and $K_p = 585, K_i = 50\,000$, $K_d = 3.55$

requirements of fast, stable response with no steady-state error. As in the case of the other types of controllers discussed so far, the gains interact with each other quite a bit. Especially, the K_p gain multiples the K_d and K_i gains. Any change made in the K_p gain will change the effective value of the K_d and K_i gains. Hence, they may need to be re-tuned, too.

TUNING A CASCADED VELOCITY/POSITION CONTROLLER WITH FEEDFOR-WARD GAINS

1. Set all gains to zero,
2. TUNING K_p AND K_d: To tune these gains we want to begin from an underdamped system response with slight damping. If the mechanical system already has some damping effect due to friction, then

Continued

TUNING A CASCADED VELOCITY/POSITION CONTROLLER WITH FEED-
FORWARD GAINS (CONTINUED)

we can begin tuning by setting $K_d = 0$. Otherwise, K_d should be set
to a low value to introduce some damping into the system by the
controller. Provide a small position step input (such as quarter to
half rotation of the motor shaft) and set up to capture the resulting
actual position. Now, increase K_p to get the fastest possible rise time
without a large overshoot. In this step, slightly more overshoot can
be allowed since it will be cut down next by the K_d gain. Once the
fastest response is obtained, start increasing K_d to increase the damping in the system. This will bring down the overshoot but will also
slow down the response. Continue to adjust these two gains until a
critically damped response or close to it can be obtained.

3. TUNING K_i: Tuning the integral gain is difficult because it can
   cause overshoot. Increase K_i as high as possible without causing any overshoot. Integrator saturation will generate a large integral signal resulting in a big overshoot or oscillations. Therefore,
   often integration is activated only when no motion is commanded
   (integrate-when-in-position). This scheme allows taking out small
   following errors after the motion stops.

4. TUNING K_{vff}: Setting the K_{vff} to the same value as the K_d is a good
   beginning point, which usually leads to acceptable results. Sometimes, it may be necessary to increase it slightly based on the settings
   of the other gains, especially the K_{aff}. To adjust this gain, position
   input coming from a trapezoidal velocity profile should be applied
   as input while capturing the actual position and the following error.
   Increase K_{vff} until the following error is reduced as low as possible
   without making the system unstable.

5. TUNING K_{aff}: To adjust this gain, position input coming from a trapezoidal velocity profile should be applied as input while capturing the
   actual position and the following error. Increase K_{aff} until the following error is reduced as low as possible or eliminated. Sometimes, it is
   necessary to go back and adjust K_{vff} and then re-tune K_{aff} to obtain
   even better results.

■ **EXAMPLE 6.5.3**

The same axis with the AC servo motor in Example 6.5.2 is now controlled by a cascaded
velocity/position controller with feedforward gains as shown in Figure 6.55. Tune the controller to obtain fast response with minimal following error and no overshoot in a point-to-point
move.

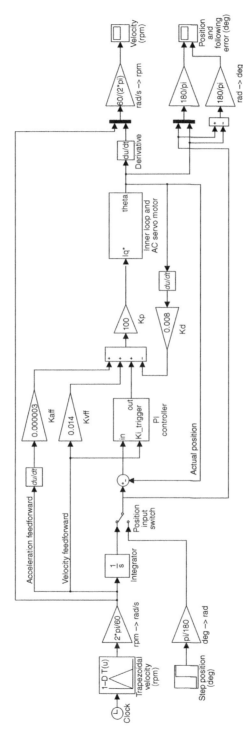

Figure 6.55 System setup for tuning the PID position controller for an AC servo axis using the Integrate-When-In-Position approach

Solution

First, all gains are set to zero. Since the system already has some damping due to friction, $K_d = 0$ was also set as a beginning point for tuning.

To tune the K_p and K_d gains, a position step of 90° $\left(\frac{1}{4} \text{turn}\right)$ was set up. The Position input switch was set to the step input and the resulting actual position was captured. The K_p gain was increased slowly until $K_p = 100$ where the response had about 20% overshoot and oscillations as shown in Figure 6.56a. Then, K_d was increased to 0.006 to obtain the fastest possible rise time with minimal overshoot as seen in Figure 6.56b.

To tune the K_i gain, we start increasing the K_i gain in small increments until some overshoot and/or oscillations are observed. Then, the K_i gain is backed down to 50 where the following error was eliminated as shown in Figure 6.56c. Note that the system uses the integrate-when-in-position method.

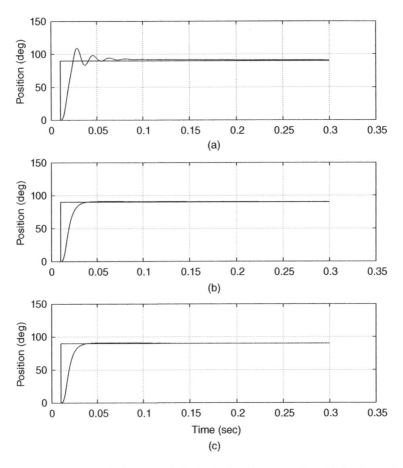

Figure 6.56 Tuning K_p, K_d gains in a cascaded velocity/position controller with feedforward. (a) $K_p = 100, K_d = 0$; (b) $K_p = 100, K_d = 0.006$; (c) $K_p = 100, K_d = 0.006, K_i = 50$

To tune the K_{vff} *gain*, first we need to change the input from a position step into a position input coming from trapezoidal velocity. Under normal operating conditions, the axis will make a point-to-point move using a trapezoidal velocity profile. In Figure 6.55 the `Position input switch` was set to the trapezoidal velocity input for 200 rpm. The resulting actual position and the following error were captured.

First, let us look at the response with just $K_p = 100, K_d = 0.006, K_i = 0$. Refering to Section 6.2.3.1, this makes the system *Type-1*, which leads to following error during the constant velocity segment of the input velocity profile as shown in Figure 6.57.

We will set $K_{vff} = K_d$ as a good starting point as described in the tuning procedure. Figure 6.58 shows the significant improvement in the following error which went from a maximum of 8.4° down to 1.2°.

To tune the K_{aff} *gain*, we increase it starting from a very small value. As the gain is incremented, the following error goes down. Figure 6.59 shows the system response with $K_{aff} = 0.000003$. Note that K_{vff} had to be increased slightly to 0.008 to achieve the sharp corners in the velocity response. The maximum following error while in motion is 1°, which

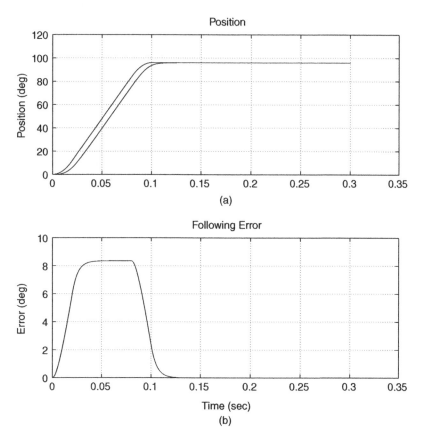

Figure 6.57 Type-1 system response using a cascaded velocity/position controller with feedforward ($K_p = 100, K_d = 0.006, K_i = 0, K_{vff} = 0$). (a) Position response; (b) following error

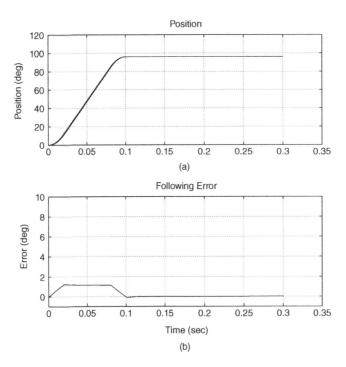

Figure 6.58 System response using a cascaded velocity/position controller with feedforward ($K_p = 100, K_d = 0.006, K_i = 0, K_{vff} = K_d$). (a) Position response; (b) following error

is almost zero in practical terms. The K_i gain is activated to integrate when in position to help reject any disturbances when in position (stopped). For example, the system rejects a sudden 8 Nm constant disturbance at $t = 0.15$ in Figure 6.59c. When the disturbance occurs, the following error increases to 3° but then the system recovers quickly and eliminates the error.

As always, while tuning the gains, one must make sure that the current command does not saturate. In this system, the maximum instantaneous current command was about 2.7 A and the continuous current was about 2 A during motion, which are well within the capabilities of a typical industrial motion controller.

■ **EXAMPLE 6.5.4**

The turret axis of the core capping machine in Example 3.11.1 in Chapter 3 is controlled by a cascaded velocity/position controller. Tune the controller so that the desired motion can be achieved.

Solution
In Example 3.11.1, a three-phase AC vector induction motor and a gearbox were selected for the machine. The motion has a triangular velocity profile to rotate the turret 180° in 0.8 s. The final total inertia on the motor shaft was calculated as $J = 0.142$ lb-ft^2.

Figure 6.60 shows the control system diagram. The Inner loop and AC induction motor block contains the vector controller for AC induction motor and the

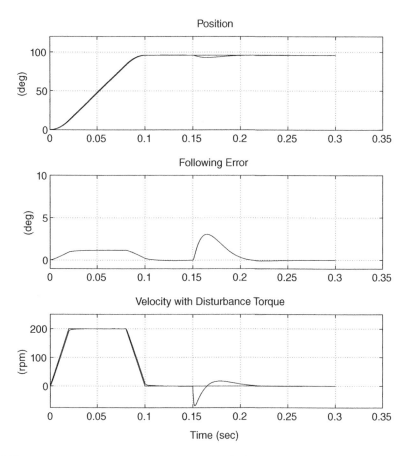

Figure 6.59 System response using a cascaded velocity/position controller with feedforward ($K_p = 100, K_d = 0.008, K_i = 50, K_{vff} = 0.008, K_{aff} = 0.000003$). A disturbance load torque of 8 Nm was applied at 0.15 s

corresponding current loops as discussed in Section 6.4.1 and Figure 6.37. In this model, all electrical properties of the selected motor, the total system inertia and friction torque of $0.23\ lb - ft$ were used as parameters for the simulation.

The triangular velocity profile in the `Triangular Velocity (rpm)` block was programmed into the simulation by finding the maximum rotational velocity ω_{max} required for the motion (peak of the triangle). The area under the triangular velocity profile equals the distance traveled by the axis. Since the axis completes half a revolution in 0.8 s

$$s = \frac{\omega_{max}0.8}{2}$$

$$\frac{1}{2} = \frac{\omega_{max}0.8}{2}$$

solving for ω_{max} gives 1.25 rev/s , which is 75 rpm.

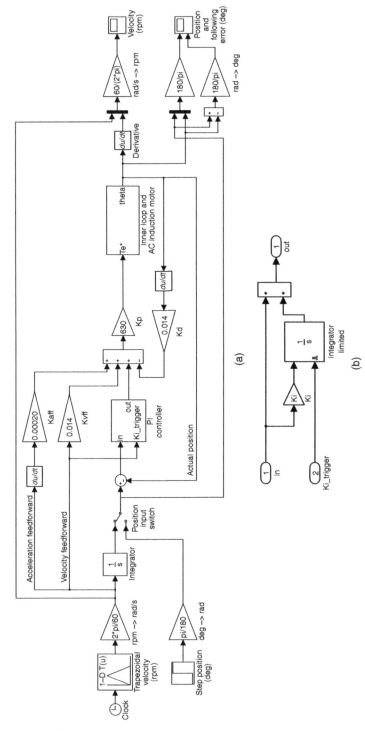

Figure 6.60 Position control of the turret axis in Example 3.11.1 with AC induction motor and gearbox. The motion controller uses cascaded velocity /position controller with feedforward. (a) Control system configuration. (b) Inside the PID block. The integrator was set up to be triggered by an external signal (commanded velocity)

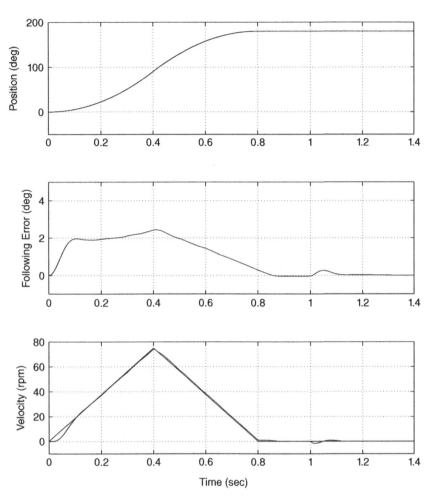

Figure 6.61 Turret axis response using a cascaded velocity/position controller with feedforward ($K_p = 630, K_d = 0.014, K_i = 25, K_{vff} = 0.014, K_{aff} = 0.00020$). A disturbance load torque of $3\,\text{lb} - \text{ft}$ was applied at 1 s

The controller was tuned using the tuning method in Section 6.5.3 and the resulting gains are shown in Figure 6.60. The turret axis can follow the triangular velocity profile while completing the 180° motion as shown in Figure 6.61. Initially, there is a relatively large following error due to the inertia of the load. After completing the motion in 0.8 s, the turret stops briefly to insert the caps. The axis was disturbed by a $3\,\text{lb} - \text{ft}$ load torque at 1 s to assess how well the controller could reject the disturbance. The integrate-when-in-position method was used in the simulation.

Problems

1. Table 6.2 shows switching patterns using the 180° conduction method with the inverter shown in Figure 6.6. Complete the transistor switching waveforms in Figure 6.62. The waveforms corresponding to the first two rows of Table 6.2 have already been shown in the figure.

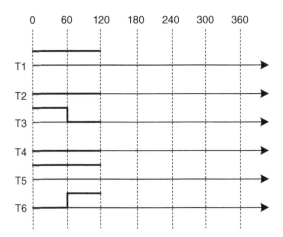

Figure 6.62   First two intervals of the transistor switching waveforms for the 180° conduction method, Problem 1

2. Referring to Table 6.2 and Figure 6.6, complete the line-to-line voltage waveforms in Figure 6.63. The waveforms for the 0°–120° interval have already been provided in the figure. For example, in the 0°–60° interval T1 is ON, which connects point "a" in the inverter leg in Figure 6.6 to the positive side of the supply voltage. Similarly, T5 is ON, which connects point "b" to the negative side of the supply voltage. As a result, the line-to-line voltage between points "a" and "b" becomes V. Verify your waveforms using the terminal voltages given in the last three columns in Table 6.2.

3. Build a Simulink® diagram similar to Figure 6.11 to create a PWM signal from a sinusoidal reference wave. The diagram in Figure 6.11 has three input waves. Yours will have only one input wave (reference sinusoidal). The DC bus voltage is 1 V (not center tapped). The carrier frequency is 20 Hz and amplitude is 1 V. The reference sinusoidal amplitude is 0.8 V with 2π rad/s frequency. Plot the reference, carrier, and PWM signals as shown in Figure 6.8. Hint: Your signals should look just like those in Figure 6.8.

4. Show that the CLTF given in Equation (6.25) is equivalent to the system shown in Figure 6.28.

5. Build a Simulink® model for the system shown in Figure 6.22b to simulate its response to unit step input. Use $K_m = 100$ and $m = 0.5$.
   (a) Plot the response with $K_d = 0.1$ and $K_p = 0.3$, $K_p = 0.6$, $K_p = 1.2$. Explain the effect of increasing the K_p gain on the response while keeping K_d constant.
   (b) Plot the response with $K_p = 5$ and $K_d = 0.1$, $K_d = 0.2$, $K_d = 0.4$. Explain the effect of increasing the K_d gain on the response while keeping K_d constant.

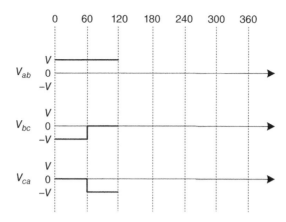

Figure 6.63 First two intervals of the line-to-line voltage waveforms for the 180° conduction method, Problem 2

   (c) Tune the controller using the basic PID tuning method to obtain a response with 10% overshoot and 1.5 s settling time. Note that $K_i = 0$.

6. Modify the Simulink® model you built in problem 5 so that the desired position input is derived from a trapezoidal velocity profile with $t_a = t_d = 100$ ms and $V_m = 4000\,\text{cts/s}$. Also, assuming the motor has an encoder that generates 8000 cts/rev, modify the diagram so that the actual position is in cts and the desired velocity input is in cts/s.

   (a) Make two plots for position (desired and actual on the same plot) and velocity profile.

   (b) What is the maximum following error in cts?

   (c) What is the position in cts when the axis moves for 1 s?

Hint: Assume that the output of the system transfer function is in radians. Insert a transfer function for the encoder after the system transfer function to convert the position into *cts*. This transfer function will be a simple conversion factor. To create the desired position input from a trapezoidal velocity profile, replace the `Step` input block with three blocks. The first two blocks will be a `Clock` and a `Lookup Table` block to define the velocity profile. The third block will be an `Integrator` to generate the desired position from the velocity profile. This is similar to the input generation in Figure 6.55.

7. Build a Simulink® model for the system in Figure 6.25. Use the integrate-when-in-position approach in the PID controller. The system transfer function is $G(s) = \frac{100}{0.5s^2}$.

   (a) Do not apply any disturbance and turn the integrator OFF by setting $K_i = 0$. Tune the K_p and K_d gains to obtain a fast response with 10% overshoot, settling time of less than 0.5 s and no steady-state error.

   (b) Keep the integrator OFF and apply step disturbance of 0.1 units at $t = 1$ second. Can the steady-state error be eliminated by increasing the K_p gain? What happens to the overshoot?

   (c) Activate the integrator and tune the K_i gain. Can the steady-state error be eliminated? What are the final values for the K_p, K_d and K_i gains?

8. Some motion controllers can apply parabolic velocity input to the motor during tuning. Since the velocity profile is parabolic (quadratic), the position profile will be cubic and

the acceleration will be varying linearly with finite maximum values. As a result, the motion is smoother since the jerk is finite as in the case of S-curve velocity profile. Build the Simulink® model in Figure 6.55 but replace the Inner loop block with a simple Transfer Fcn block where $G(s) = \frac{100}{0.5s^2}$. Also, replace the trapezoidal velocity profile with a parabolic velocity profile generator. You need to build the parabolic velocity profile generator using blocks from the Simulink® library such that its output looks like Figure 6.64.

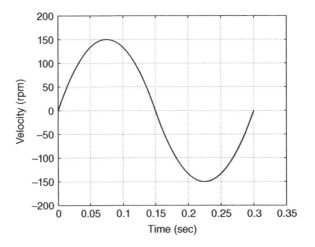

Figure 6.64   Parabolic velocity input, Problem 8

(a) Switch to step input (with 3 rad amplitude) and tune the K_p and K_d gains to obtain a fast response with no overshoot. Make sure that all other gains are set to zero. Plot the step response.
(b) Switch to parabolic velocity input with the same K_p, K_d gains. Plot the position, velocity and following error. What is the maximum following error (rad)?
(c) Tune the K_{vff} and K_{aff} gains. Plot the position, velocity and following error. What is the maximum following error (rad)?
(d) Add a step disturbance to the diagram similar to the one in Figure 6.25. Apply a disturbance of 5 units at $t = 0.15$ s. Tune the K_i gain. Plot the position, velocity and following error. What is the maximum following error (*rad*)?

9. The motion controller for the flying saw machine in Example 3.8.1 in Chapter 3 uses cascaded position/velocity controller with feedforward gains as in Example 6.5.3 and Figure 6.55.

(a) Build a Simulink® model as in Figure 6.55 and all of its blocks and subsystems explained in the book. The gains for the current loop should be set to $K_p = 100$ and $K_d = 500$. The PWM inverter has 340 VDC bus voltage, 15 VDC carrier signal amplitude and 10 kHz carrier frequency. Inertia J in the mechanical model embedded in the AC Servo Motor subsystem block of the diagram should be set equal to the total inertia seen by the motor ($J_{\text{total}} = J_{\text{ref}}^{\text{trans}} + J_m$). This will allow the simulation to account for the motor and the belt drive transmission.

(b) Tune the controller so that the saw follows the desired velocity profile described in Example 3.8.1 with less than 0.1 in following error. What are the gain settings for $K_p, K_d, K_i, K_{\text{vff}}$ and K_{aff} for the cascaded loop controller?

(c) What is the following error at the instant the saw returns to its original position ($t = 3.5\,\text{s}$)?

10. The motion controller for the glue machine in Example 3.12.1 in Chapter 3 uses cascaded position/velocity controller with feedforward gains as in Example 6.5.3 and Figure 6.55.

(a) Build a Simulink® model as in Figure 6.55 and all of its blocks and subsystems explained in the book. The gains for the current loop should be set to $K_p = 100$ and $K_d = 500$. The PWM inverter has 340 VDC bus voltage, 15 VDC carrier signal amplitude and 10 kHz carrier frequency. Inertia J in the mechanical model embedded in the AC Servo Motor subsystem block of the diagram should be set equal to the total inertia seen by the motor (J_{total}). This will allow the simulation to account for the motor, gearhead and the belt drive transmission. From the datasheet of the selected motor $R = 2.2\,\Omega, L = 13$ mH and $p = 8$ poles.

(b) Tune the controller so that the glue head follows the desired velocity profile described in Example 3.12.1 with less than 2 mm following error. What are the gain settings for $K_p, K_d, K_i, K_{\text{vff}}$ and K_{aff} for the cascaded loop controller?

(c) What are the maximum following error and the final position error when the glue head stops?

References

[1] Three Phase Bridge VS-26MT.., VS36MT.. Series (2012). http://www.vishay.com/docs/93565/vs-36mtseries.pdf (accessed 6 November 2014).

[2] IGBT SIP Module, CPV362M4FPbF (2013). http://www.vishay.com/docs/94361/cpv362m4.pdf (accessed 6 November 2014).

[3] Bimal K. Bose. *Modern Power Electronics and AC Drives*. Prentice-Hall PTR, 2001.

[4] Sabri Cetinkunt. *Mechatronics*. John Wiley & Sons, Inc., 2006.

[5] Mohamed El-Sharkawi. *Fundamentals of Electric Drives*. Brooks/Cole, 2000.

[6] Mohamed El-Sharkawi. *Electric Energy, An Introduction*. CRC Press, Third edition, 2013.

[7] George Ellis. Twenty Minute Tune-up: Put a PID on It, 2000.

[8] George Ellis. *Control Systems Design Guide*. Elsevier Academic Press, Third edition, 2004.

[9] Marvin J. Fisher. *Power Electronics*. PWS-KENT Publishing Co., 1991.

[10] Paul Horowitz and Winfield Hill. *The Art of Electronics*. Cambridge University Press, 1996.

[11] Takashi Kenjo. *Electric Motors and Their Controls, an Introduction*. Oxford University Press, 1991.

[12] Prabha S. Kundur. *Power System Stability and Control*. McGraw-Hill, Inc., 1994.

[13] Norman Nise. *Control Systems Engineering*. John Wiley & Sons, Inc., 2007.

[14] M. Riaz (2011). Simulation of Electrical Machine and Drive Systems Using MATLAB and SIMULINK. http://umn.edu/ riaz (accessed 7 April 2012).

[15] Graham J. Rogers, John D. Manno, and Robert T.H. Alden. An Aggregated Induction Motor Model for Industrial Plant. *IEEE Transactions on Power Apparatus and Systems*, PAS-103(4):683–690, 1984.

[16] Charles Rollman. What is 'Field Oriented Control' and What Good Is It. Technical report, Copley Controls Corp., 2002.

[17] Siemens Industry, Inc. (2014) Basics of AC Drives. http://cmsapps.sea.siemens.com/step/flash/STEPACDrives/ (accessed 12 November 2014).

[18] Bin Wu. *High-Power Converters and AC Drives*. IEEE Press, A John Wiley & Sons, Inc. Publication, 2006.

[19] Amirnazer Yazdani and Reza Iravani. *Voltage-Source Converters in Power Systems Modeling, Control and Applications*. IEEE Press, A John Wiley & Sons, Inc. Publication, 2010.

7

Motion Controller Programming and Applications

A motion controller is a programmable device and is the "brains" of the system. It generates motion profiles for all axes, monitors I/O, and closes feedback loops. The controller can also generate and manage complex motion profiles including electronic camming, linear interpolation, circular interpolation, contouring, and master/slave coordination.

The chapter begins by exploring linear, circular, and contour move modes. Controllers can mimic the functionality of a typical programmable logic controller (PLC) hardware. A PLC is another type of controller which is widely used in automation. The chapter continues by introducing algorithms for basic PLC functionality that are commonly used in motion controller programs. Next, jogging and homing moves are presented, which are single-axis moves. Following this, algorithms for multiaxis motion with coordinated moves, master/slave synchronization, electronic camming, and tension control are presented. Non-Cartesian machines, such as industrial robots can be controlled by motion controllers. However, the complex geometry of the machine requires kinematic calculations in real time. The chapter concludes by reviewing forward and inverse kinematics and by providing these routines for a selective compliance articulated robot arm (SCARA) robot.

Each controller manufacturer has its own programming language and programming environment for their products. Since the programming details are very specific to each hardware, only algorithmic outlines are provided in this chapter. The product manuals and manufacturer suggestions must be closely followed in adapting these algorithms to a specific choice of motion controller hardware.

7.1 Move Modes

The main function of a motion controller is to make axis position and trajectory moves. The controller can make different types of moves, such as linear or circular, between two points. Once the move is commanded, the trajectory generator creates intermediate positions (points) between the start and finish in real time. These positions are then passed on to the control system to manage the motion of each axis.

Industrial Motion Control: Motor Selection, Drives, Controller Tuning, Applications, First Edition. Hakan Gürocak.
© 2016 John Wiley & Sons, Ltd. Published 2016 by John Wiley & Sons, Ltd.

Each move consists of a move type and constraints, which are move position/distance, maximum velocity, and acceleration/deceleration. Based on this information, the trajectory generator creates a velocity profile, which is either trapezoidal or S-curve as discussed in Chapter 2. Following this velocity profile, the axis begins its move from the starting point, accelerates to the specified speed, runs for a while, decelerates, and comes to a stop at the finish point.

7.1.1   Linear Moves

This is the most basic move mode that executes along a straight line between two points. It requires the start and end point as well as the move constraints.

The trapezoidal velocity profile requires acceleration/deceleration times (t_a) and the move time (t_m) as shown in Figure 2.4. If these times are specified in the user program, then the controller computes the maximum move velocity (V_m). If the user program specifies the maximum move velocity, often as *feed rate*, then the controller computes t_m and t_a to generate the velocity profile.

There are two ways to specify the target positions: (1) *Absolute* and (2) *Incremental*. Let us assume that two consecutive moves will be made for 5 and 10 in with the axis starting at an origin. In the absolute mode, each move is made with respect to the origin location. Therefore, in the first move the axis travels 5 in. In the second move, it travels another 5 in since this would be equal to 10 in from the origin. In case of the incremental mode, the subsequent move is made from the current position. Hence, in this example, the axis would first move 5 in followed by 10 in stopping at 15 in away from the origin.

7.1.2   Circular Moves

In this move mode, a circular arc is generated between two points as the path to be followed by a tool tip (such as a cutting tool). The circle can be contained in a two-dimensional (2D) plane in any orientation or it can be a three-dimensional (3D) arc as in the case of a helix. This mode requires multiple axes to be coordinated by the controller automatically.

The circular move commands require definition of the orientation of the plane that contains the arc. This is usually done by specifying the coordinates of the normal vector for the plane. Moreover, the start point, end point, and center point of the circular arc need to be specified. Since there can be a clockwise (CW) or counterclockwise (CCW) arc between two points in the same plane, the direction of the move must also be specified (Figure 7.1). Finally, the usual motion constraints (maximum velocity, acceleration/deceleration) are needed for the trajectory generator to create the move profile.

7.1.3   Contour Moves

In some applications, such as glue dispensing, computer numerical control (CNC), welding or scanning, the trajectory is made up of many segments. Contour moves are necessary when a trajectory cannot be constructed by using line and arc segments. In this case, a sequence of positions are specified by the user program. The controller connects these points using

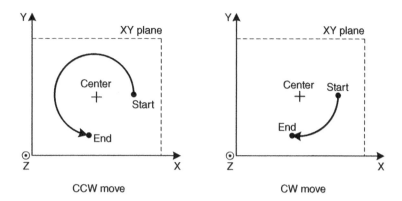

Figure 7.1 Circular moves in the *XY* plane

splines to generate the trajectory. Typically, position profile between two adjacent points is generated using cubic splines. This prevents sudden changes in velocity or acceleration at the spline boundaries between adjacent segments. The total travel time between the first and last position is split into equal time segments to execute each position segment. Blending between two linear segments, two circular segments, and between a linear and circular segment are possible. Blending enables continuous smooth motion along the entire trajectory even though it consists of segments.

7.1.3.1 Velocity Blending

The controller can smoothly transit from one velocity to another by blending the velocity of the first move to the velocity of the next move. The blending starts at the point where the deceleration for the first move begins as shown in Figure 7.2a. When blending is activated, the end position of each segment may or may not be reached.

As shown in Figure 7.2b, when blending is activated, the tool tip will start from point "A" travel toward "B" and smoothly and continuously change its direction toward "C" and come to a stop at "C." But point "B" will not be reached. If the blending is deactivated, the tool tip will start from "A" move to "B" and stop momentarily. Then, it will accelerate toward "C" continue to move, and finally stop at "C."

7.2 Programming

A motion controller program consists of instructions to describe the desired motion and the general control logic for the machine. The logic control is implemented as a PLC program. Each motion controller manufacturer has its own approach for programming the controller. In some controllers, the motion commands are intermixed with the PLC programs [18, 19] and in others, the motion and PLC programs are written as separate programs that coordinate to execute the desired motion [12]. In the following sections, we will look at the case of the separate programs.

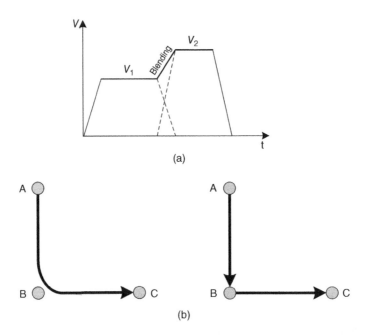

Figure 7.2   Blended moves. (a) Blending of velocities between two consecutive moves. (b) Two linear moves (Left) with and (Right) without blending

7.2.1   Motion Programs

All functionality of the motion controller describing the desired motion can be programmed using mathematical, logical, machine I/O operations, and commands for specific types of moves. Controller parameters, such as gains or any configuration parameters, can be set from the program. A motion controller can store many programs in its memory. One motion program can call other motion programs as subroutines.

Each motion controller manufacturer has its own programming language syntax. However, all of these languages have constructs to control the logic flow in a program. Typical constructs include WHILE loops, FOR loops and IF . . . THEN branching. In addition to the basic mathematical and logical operators, controllers with kinematics capabilities come with additional mathematical features such as SIN, COS, SQRT, etc., Algorithm 1 shows the basic structure of a typical motion program.

7.2.2   PLC Functionality

A PLC is a special-purpose industrial computer. PLCs have been the most widely used technology in automation since 1960s. A PLC program is often written in ladder logic format [17] and runs continuously in a so-called *scan mode*. Essentially, the program runs in an *infinite loop* once started. In each scan cycle, the PLC first reads the state of all I/O devices. Then, it scans the logic in the program from top to bottom to determine the new states for the output devices given the current states of all input devices. Finally, it updates the output devices.

Algorithm 1 Structure of a basic motion program

1: **procedure** MAIN
    Assign motors to axes
2:     $X \leftarrow Motor1$ ▷ Motor 1 assigned to X axis
3:     $Y \leftarrow Motor2$ ▷ Motor 2 assigned to Y axis
    Define move modes
4:     $MoveMode \leftarrow Linear$ ▷ Linear moves
5:     $PosMode \leftarrow Abs$ ▷ Absolute positions
    Define move constraints
6:     $TA \leftarrow 10$ ▷ 10 ms accel time
7:     $TM \leftarrow 1000$ ▷ 1000 ms travel time
    Desired motion
8:     $X5\ Y3$ ▷ Move 5 units in X and 3 units in Y
9: **end procedure**

Depending on the complexity of the logic program and the speed of the CPU, each scan cycle may take only a few milliseconds per thousand lines of simple logic statements.

Motion controllers mimic the scan cycle functionality of a typical PLC hardware as a software that runs on the motion controller itself. The motion controller can run PLC programs in the background while also executing the motion command/programs. The PLC programs are ideally suited for tasks that are *asynchronous* to the sequenced motion. They can contain the same syntax as a motion program except for motion commands. Typically, PLC programs implement monitoring inputs, updating outputs, changing controller gains, configuring hardware, sending commands, and communication with the human–machine interface (HMI).

In the following sections, we will explore a few basic sample PLC programs that can be used to implement various functionality often encountered in motion control applications.

7.2.2.1 Reading Inputs

Machine I/O is mapped to the memory of the controller. For example, let us assume that there is a D-sub style 25-pin connector on the controller for its digital inputs and outputs. This is where ON/OFF type user interface devices, such as operator buttons or indicator lights, can be wired. Let us also assume that there are 12 input pins and 12 output pins in this connector.

In the memory-mapped approach, all input pins of the controller can be mapped to a 12-bit wide location in the memory of the controller. Each physical pin in the connector corresponds to a single bit in this 12-bit wide memory location. Similarly, the 12 output pins can be mapped to another 12-bit wide location in the memory. These memory locations can be accessed from the user's PLC program by assigning variable names to each bit. Then, by reading the value of an input variable, the program can gather the state of that input device. Similarly, by equating the output variable of an output device to an appropriate value, the program can turn on or off the external output device wired to that pin.

Virtually, all automatic machines have push buttons similar to those in Figure 5.22 on their control panels. These devices are wired to the digital inputs. Assume that a push button is wired to input pin 1 and *INP*1 is assigned to that pin (bit) as the input variable in the software.

The input signal turns on when the button is pressed. This corresponds to logic high (= 1) in the software. The signal turns off when it is released, which is logic low (= 0). Algorithm 2 shows a PLC building block that can capture the state of this button and activate/deactivate an indicator light (Figure 5.26) wired to output bit 10 assigned to variable $OUT10$.

Algorithm 2 PLC Sample Program: *Reading level-triggered input* [12]

1:	**procedure** PLC	
2:	**if** ($INP1 = 1$) **then**	▷ If input bit 1 is high
3:	$OUT10 \leftarrow 1$	▷ Turn on output bit 10
4:	**else**	
5:	$OUT10 \leftarrow 0$	▷ Turn off output bit 10
6:	**end if**	
7:	**end procedure**	

7.2.2.2 Edge-Triggered Output

Sometimes it is necessary to take an action only once on the rising edge of an input signal. The *rising edge* is the transition from off to on state in the input signal.

Assume that the PLC will send a command to the HMI to display a message on its screen when an input device turns on. If the simple logic in Algorithm 2 was used by replacing the $OUT10$ output with the display command, the HMI would receive several hundred commands per second from the PLC since it will have scanned the code many times as the user pressed the button. This would overwhelm the communication buffers of the controller and the HMI.

Algorithm 3 shows how we can capture the rising edge of the input signal and send the display command (CMD) only once. $LAT1$ and $INP1$ are an internal variable and the input pin variable, respectively. $LAT1$ is used as a software latch to capture the state of the input.

Algorithm 3 PLC Sample Program: *Edge-triggered output* [12]

	Initialize	
1:	$LAT1 \leftarrow INP1$	▷ Set initial input state to latch state
2:	**procedure** PLC	
3:	**if** ($INP1 = 1$) **then**	▷ If input 1 is high
4:	**if** ($LAT1 = 0$) **then**	▷ If input 1 was previously low
5:	CMD *"display"*	▷ Send a display command to the HMI
6:	$LAT1 \leftarrow 1$	▷ Set the latch
7:	**end if**	
8:	**else**	▷ Input 1 is low
9:	$LAT1 \leftarrow 0$	▷ Reset the latch
10:	**end if**	
11:	**end procedure**	

7.2.2.3 Precise Timing

Motion controllers have built-in timers. These timers can be used to create a precise time delay during the operation of the machine such as to blink a light every 2 s.

Timer ticks come from a constant frequency reference clock or the servo cycle. In each servo cycle, the timer receives one tick and is updated by either counting up or down. If the servo cycle is t_{srv} s long, then for a delay of dly s, the timer will have to count down from dly/t_{srv}. For example, if $t_{srv} = 0.001$ s, for a 2 s delay, the timer will have to count down from $2/0.001 = 2000$.

Assume that the controller has a timer $T1$ that counts down from a preset value to zero by t_{srv} amount in every servo cycle tick. Algorithm 4 shows how to create a delay for precisely dly seconds.

Algorithm 4 PLC Sample Program: *Precise timing*

```
1: procedure PLC
2:     T1 ← dly/tsrv                          ▷ Load the total count into timer
3:     while (T1 > 0) do                                      ▷ Loop until zero
4:     end while
5: end procedure
```

7.2.2.4 Reading Selector Switch

Selector switches are used to choose between different actions. Assume that a machine has a *3-position selector switch* (maintained) as shown in Figure 5.25. The switch has *left*, *center*, and *right* positions and is wired to three input terminals. In these switches, different types of cams can be used leading to various signal combinations between the inputs to indicate each of the three switch positions. In this example, the switch cam is such that in each position one of the inputs is high while the others are low.

Let us assume that a motor needs to be run in positive or negative direction or stopped using the 3-position selector switch. Algorithm 5 shows how to detect the switch state and show the corresponding motor state on an HMI. *LAT1*, *LAT2*, and *LAT3* are internal variables. *INP1*, *INP2*, and *INP3* are input terminal variables for each of the switch positions. Latching is used to detect the edge transitions in the input signal to avoid sending the same command repeatedly to the HMI due to the scan cycle of the PLC.

7.3 Single-Axis Motion

As the name implies, single-axis motion is about moving one axis at a time without any coordination with any other axes. There are two main types of single-axis moves that are extensively used in motion control industry: (1) Jogging and (2) Homing. In these moves, trapezoidal or S-curve velocity profiles are used in the trajectory computations.

Algorithm 5 PLC Sample Program: *Selector switch*

Initialize
1: $LAT1 \leftarrow 0; LAT2 \leftarrow 0; LAT3 \leftarrow 0$
2: **procedure** PLC
3:　　**if** ($INP1 = 1$ and $INP2 = 0$ and $INP3 = 0$ and $LAT1 = 0$) **then**　　　　▷ SW: left
4:　　　　CMD "*Motor Positive*"
5:　　　　$LAT1 \leftarrow 1; LAT2 \leftarrow 0; LAT3 \leftarrow 0$　　　　　　　　　　　　▷ latches
6:　　**else if** ($INP1 = 0$ and $INP2 = 1$ and $INP3 = 0$ and $LAT2 = 0$) **then**
　　　　　　　　　　　　　　　　　　　　　　　　　　　　　　　　　　　▷ SW: center
7:　　　　CMD "*Motor Stop*"
8:　　　　$LAT1 \leftarrow 0; LAT2 \leftarrow 1; LAT3 \leftarrow 0$　　　　　　　　　　　　▷ latches
9:　　**else if** ($INP1 = 0$ and $INP2 = 0$ and $INP3 = 1$ and $LAT3 = 0$) **then**
　　　　　　　　　　　　　　　　　　　　　　　　　　　　　　　　　　　▷ SW: right
10:　　　　CMD "*Motor Negative*"
11:　　　　$LAT1 \leftarrow 0; LAT2 \leftarrow 0; LAT3 \leftarrow 1$　　　　　　　　　　　　▷ latches
12:　　**end if**
13: **end procedure**

Three modes of operation are commonly used to jog or home an axis. First, jogging commands can be issued from an operator terminal (host computer) to the motion controller. The controller receives the command, computes the trajectories, and executes the motion. Second, a user interface with buttons can be used. For example, when the jog button is pressed for axis 1 and held, the axis moves until the button is released. Third, the jogging and homing commands can be used in a motion program to move the axis programmatically.

7.3.1　Jogging

Jogging is simply about moving a single axis. Setup parameters can be configured to define the speed and acceleration of the jogging motion. Commands can be issued to move in the positive or negative direction to either a specific position, through a certain distance or continuously without a predefined stop.

7.3.2　Homing

Motion controllers have built-in routines for homing search. The goal of homing is to establish an absolute reference position for an axis. This reference position is called the *home position*. Homing is especially needed when incremental encoder position feedback is used since upon power-up, the axis position will be unknown. All subsequent moves are defined with respect to the home position for the axis once it is found.

Although there are several methods for homing, the most common method uses a *move-until-trigger* scheme to find the home position established by a sensor. Figure 7.3 shows a linear axis with two limit switches at the ends of travel and a home switch in the middle. Typically, these are magnetic switches that are wired to the corresponding digital I/O for that axis on the controller.

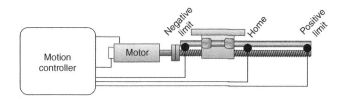

Figure 7.3   Linear axis with two limit switches and a home switch

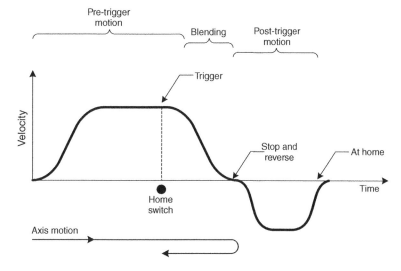

Figure 7.4   Homing search based on the move-until-trigger scheme (Adapted by permission of Delta Tau Data Systems, Inc. [12])

The move-until-trigger scheme consists of a *pre-trigger* and *post-trigger* segment (Figure 7.4). In the pre-trigger segment, the axis starts moving toward the home sensor with a desired homing speed and acceleration. When the axis hits the home switch, a trigger is received and the pre-trigger motion begins to decelerate as the axis is now on the other side of the home switch. The controller blends this motion into the post-trigger motion. As a result, the axis smoothly changes its direction and moves back toward the home switch.

At the instant the trigger is received, the position of the axis (encoder counts) is saved using the hardware capture feature of the controller. This is an accurate way of capturing the instantaneous position of the axis when the trigger occurs. The post-trigger motion is executed for a prespecified distance from this trigger position.

Using a combination of the home switch signal with the motor encoder index channel (CH C) pulse can result in more accurate homing. As shown in Figure 7.5, the hardware capture feature of the controller can be set up to capture the rising edge of the first encoder index channel pulse after the home switch was triggered. After this trigger, the axis slows down to a momentary stop before beginning its motion back toward the home switch (Figure 7.4). At this point, the post-trigger distance between the stop and the trigger location is known very

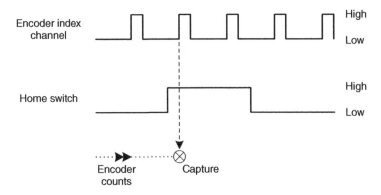

Figure 7.5   Triggering with the home switch and encoder index channel pulse to capture a more accurate home position

accurately. Therefore, the controller can compute a new trajectory to move the axis back to the trigger position, which is recorded as the home position.

A *home offset* distance can be defined in the software. This allows setup adjustments in the software for the location of the home sensor as opposed to physically adjusting the location of the sensor on the machine. This offset is added to the move from the trigger point.

7.4   Multiaxis Motion

Synchronized or coordinated multiaxis motion can involve:

1. Multiple motors driving one axis
2. Coordinated motion of two or more axes
3. Following using master/follower synchronization
4. Tension control, or
5. Kinematics.

There are many motion controllers on the market that can provide most or all of these types of synchronized motion capabilities.

7.4.1   Multiple Motors Driving One Axis

In some systems, more than one motor drives a single axis. A typical example is a gantry machine shown in Figure 7.6 found in such applications as waterjet cutting, welding, engraving, or pick-and-place. In this system, two motors (1 and 2) drive the linear motion of the base axis of the machine. These motors must be synchronized to prevent skewing the base axis.

All motors of an axis are assigned to the same coordinate system (group) that defines the axis. In case of the gantry in Figure 7.6a, motors 1 and 2 would be assigned to the base axis of the system. This way, when the motion controller commands a move for the base axis, the same set point computed by the trajectory generator is commanded to the servo loops of both

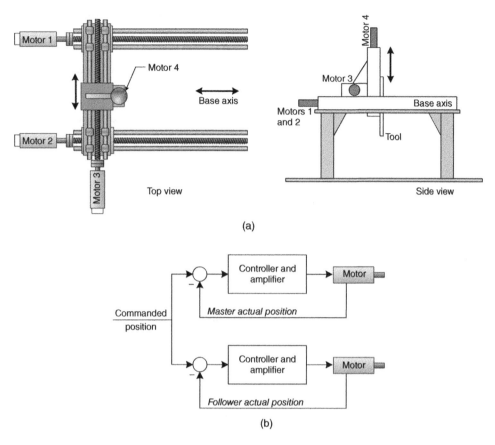

Figure 7.6   Gantry machine with set-point coordinated base axis. (a) Gantry machine with two motors driving the base axis. (b) Same set-point commanded to both motors of the base axis

motors (Figure 7.6b). Each servo loop must then track the commanded motion. Assuming all servo loops are well tuned, this method can produce a well-coordinated motion between the motors driving the same axis. This type of coordination is also called *set-point coordinated motion*. Alternatively, the two motors can be coordinated by master/follower synchronization explained in Section 7.4.3.1.

7.4.2   Coordinated Motion of Two or More Axes

When multiple axes are controlled by the motion controller, they can be coordinated to create complex motion. One way to achieve coordination is through the circular and contour moves explained in Sections 7.1.2 and 7.1.3, respectively.

   Linear moves of two or more axes can also be coordinated. In this mode, each motor is driving a single axis but all axes move in coordination. Algorithm 6 shows coordinated linear motions of X and Y axes to follow a right triangle with the tool tip. The coordinated moves are

---

Algorithm 6 Multiaxis motion: *Coordinated linear motion*

```
 1: procedure MAIN
      Setup parameters
 2:      X ← Motor1                                      ▷ Motor 1 assigned to X axis
 3:      Y ← Motor2                                      ▷ Motor 2 assigned to Y axis
 4:      TA ← 100                                           ▷ 100 ms accel time
 5:      TM ← 900                                           ▷ 900 ms move time
 6:      MoveMode ← Linear                                      ▷ Linear moves
 7:      PosMode ← Abs                                     ▷ Absolute positions
 8:      X0 Y0                                   ▷ Move to 0,0 (start position)
 9:      Dwell ← 200                                        ▷ Dwell 200 ms
      Draw a right triangle with 5 units each side
10:      X5 Y0                                                ▷ Move to 5,0
11:      X0 Y5                                       ▷ Move to 0,5 (diagonally)
12:      X0 Y0                                                ▷ Move to 0,0
13: end procedure
```

---

accomplished by assigning the axes to the same coordinate system (group) and commanding their motions together in the same line as shown in lines 8 and 10 through 12.

First, motors 1 and 2 are assigned to the X and Y axes of the coordinate system, respectively. Next, the acceleration and move times are defined along with the linear move mode and absolute position coordinates. The tool tip is commanded to go to position (0,0). After waiting there for 200 ms, the tool tip is commanded to draw a right triangle with 5 units each side. Notice that in each of the lines 10 through 12, simple position coordinates are specified for the corners of the triangle. The motion controller coordinates the two axes to get to these positions automatically. In line 11, the tool tip is commanded to move along the diagonal from current position (5,0) to (0,5). Since the diagonal distance is longer than the sides of the square, the controller adjusts the speeds of the individual axes so that the tool tip still completes its diagonal motion in the specified move time (or feedrate). Both axes start and stop their motion at the same time.

7.4.3   Following Using Master/Slave Synchronization

Many motion control applications involve axes that are *not* under the control of the motion controller. For example, cutting a continuous web material, such as paper, into fixed lengths involves a feed axis which is typically not under the control of the motion controller for the cutter. The continuous feed axis may be powered by an induction motor and its variable frequency drive. But the motion of the cutter must be synchronized to the speed/position of the feed axis. Since the motion controller does not control both axes, it cannot synchronize them using coordinated motion. Another method is needed.

Synchronizing to the motion of an *external* axis is called *following*. It is also called *master/slave* configuration. The external axis is the *master* and its motion is measured usually by an encoder. The following axis is called the *slave*. Note that more than one axis can be slaved to a single master axis.

The stream of data coming from the *master encoder* is fed to the slave axis as a series of commanded positions. The slave axis then follows this trajectory. In other words, commanded positions for the slave axis come from the external encoder rather than the internal trajectory generator.

Whenever possible, coordinated motion of the axes should be used rather than the master/slave strategy. This is because the trajectories for the axes in coordinated motion are generated mathematically. They are smoother than the trajectories generated based on master encoder data as this signal can be noisy. Smooth trajectories allow higher gains for tighter control, which leads to higher performance [2].

7.4.3.1 Electronic Gearing

A *constant* gear ratio between the motion of a master axis and the slave axis can be established through electronic gearing. As shown in Figure 1.2 and explained in the beginning of this chapter, there are no physical gears between these axes. Instead, each axis has its own motor and the synchronization is implemented in software. For example, if a gear ratio of 1:5 (master:follower) is programmed, the follower axis will move five units when the master moves one unit. Electronic gearing is also called *position following* since the position of the master is followed by the position of the follower.

The gear ratio can be changed on-the-fly since it is implemented in software. This is the most notable advantage of electronic gearing. However, loading the new gear ratio into the memory of the controller can take some time which is not desirable in very time critical moves. The accuracy of the electronic gearing is strongly influenced by how well the axes were tuned. Following errors can adversely affect the synchronization of the system. Another adverse effect can come from noisy master encoder data as the slave axis will track that data and even amplify it.

A typical example for electronic gearing is a gantry machine shown in Figure 7.6a. If the master/slave approach is used, then one of the base axis motors is chosen as the master (e.g., motor 1). This motor executes the commanded trajectories generated by the controller. The second motor (motor 2) is slaved to the feedback encoder of the first motor (motor 1) for synchronization between them to prevent skewing the base. In other words, the feedback encoder of the master motor becomes the input command for the follower motor (Figure 7.7a). If the master axis is given a jog command, the slave axis simply follows it.

The trajectory tracking performance can be limited since the actual trajectory of the master motor becomes the commanded trajectory of the slave motor. Inevitably, the actual trajectory of the master may have some minor deviations from its commanded trajectory. When the slave receives the actual master motion as a command, its servo loop may add even more deviations from the desired commanded trajectory. Some controllers allow command position following in the master/slave mode with a predefined gear ratio as shown in Figure 7.7b. But this is possible if the master axis is also under the control of the same controller as the slave axis.

7.4.3.2 Electronic Camming

Consider the mechanical cam-and-follower mechanism shown in Figure 7.8. The shape of the cam determines the motion of the follower. Therefore, the linear position of the follower (slave) is a function of the angular position of the cam (master).

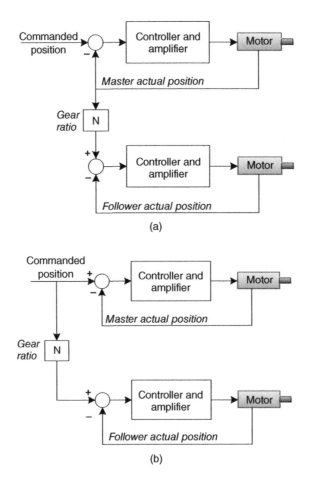

(a)

(b)

Figure 7.7 Two types of master/slave programming. (a) Master encoder following: Feedback encoder of the master motor is the command input for the slave motor through a gear ratio. (b) Command position following: Commanded master position is also sent to the slave through a gear ratio

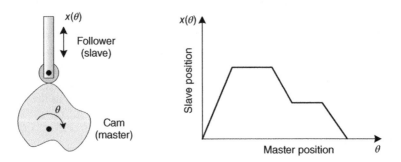

Figure 7.8 Mechanical cam-and-follower mechanism

Since the shape of the cam is nonuniform, the ratio of the follower position to the cam position is *variable*. One possibility is to use electronic gearing between the master and slave axes and change the gear ratio on-the-fly. However, the delay in loading the new gear ratio into the memory can cause the following errors.

Instead, the variable ratio can be implemented in the motion controller software using electronic camming where the change in the ratio is instantaneous. The cam can be defined in the motion software using ratio following [5], a mathematical equation, a look-up table [4], or time-base control [12]. The important thing to notice is that the slave position is programmed as a function of the master position, *not time*. This is how the two axes are locked together for synchronization.

Even though the concept originates from the mechanical cam-and-follower mechanism, the ability to implement a variable ratio in software goes much beyond just replacing the mechanical cam-and-follower with its electronic equivalent. Many sophisticated applications of multiaxis motion are implemented using electronic camming.

Ratio Following [5]

Flying knife (or *cut-to-length*) is one of the common applications of electronic camming. As shown in Figure 7.9, a set of nip rolls feeds a continuous material at a constant speed through the machine. Often this feeding axis is driven by an induction motor with its own controller and as such is not under the control of the motion controller for the machine.

A master encoder attached to the nip rolls provides position and velocity information to the motion controller to coordinate the slave axis that carries the knife. To be able to cut the moving material, the speed of the slave axis must match the speed of the material during the cutting operation. First, this requires accelerating the slave axis to the material speed, then traveling at that speed while performing the cut. Next, the slave axis must be decelerated to a stop and rapidly moved back to the beginning position of the knife for the subsequent cutting cycle to begin.

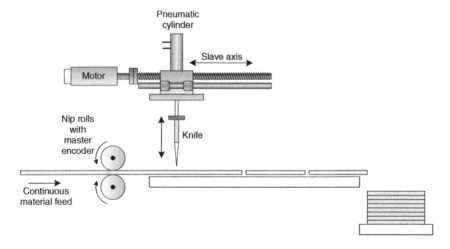

Figure 7.9   Flying knife application to cut continuous material into fixed lengths

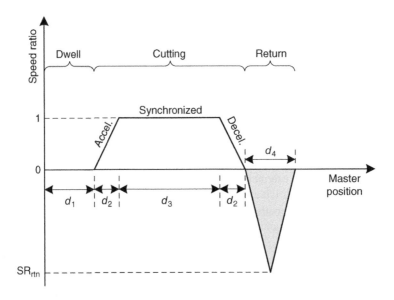

Figure 7.10   Speed ratio versus master position for synchronized motion of a flying knife (Adapted by permission of ABB Corp. [5])

One way to describe the synchronized motion is to plot the *speed ratio* against the master distance. Figure 7.10 shows the synchronized motion requirements for the flying knife application where [5, 14]

$$\text{speed ratio (SR)} = \frac{\text{slave speed}}{\text{master speed}} = \frac{V_s}{V_m} \tag{7.1}$$

The motion can be separated into three phases: (1) dwell, (2) cutting phase, and (3) return phase. The first segment of the plot is the *dwell phase*, which allows us to increase the cut length for a variable amount. During the dwell, the material travels for d_1 length. Then, the *cutting phase* starts with the acceleration of the knife to the material speed while the material travels another d_2 length. When the knife speed matches the material speed ($SR = 1$) at the beginning of the synchronized segment, a pneumatic cylinder lowers the knife into the material to cut it. The knife and the material travel through d_3 length by the time the cutting is finished. At the end of the synchronized segment, the cylinder is deactivated to retract the knife and the slave axis begins to decelerate to a stop through another d_2 length. At this point, the cutting phase is finished and the knife must begin the *return phase* to go back to its original position. A triangular velocity profile is used to rapidly move the knife back.

The area under each segment of this motion profile is the distance traveled by the slave axis. Therefore, the total area under the triangular velocity profile for the return phase must be equal to the total area under the trapezoidal velocity profile for the cutting phase so that the slave axis returns back to its exact original position

$$\frac{1}{2}d_4\, SR_{\text{rtn}} = 2\frac{1}{2}(d_2 \cdot 1) + (d_3 \cdot 1) \tag{7.2}$$

Since the speed ratio is variable, this application can be programmed using electronic camming. In the synchronized segment, *speed ratio* = 1 which means that the slave axis is traveling at the same speed as the material. Hence, the moving material appears stationary as observed from the slave axis.

Some motion controllers implement electronic camming using *ratio following* [14, 9]. Figure 7.11 shows a typical constant acceleration segment in a speed ratio versus master position motion profile. The area of the triangle is equal to

$$\text{Area} = \frac{1}{2} \left(\frac{V_s}{V_m} d_m \right) \tag{7.3}$$

where V_m and V_s are the master and slave axis velocities, respectively and d_m is the distance traveled by the master during the slave axis acceleration. Since the master is moving at constant speed

$$d_m = V_m t \tag{7.4}$$

substituting (7.4) into (7.3) gives

$$d_s = \text{Area} = \frac{1}{2}(V_s t)$$

which is the distance traveled by the slave axis d_s as the master travels d_m. In each segment of the motion, the elapsed time t is the same for both the master and the slave axes. Since the master is moving at constant speed, the distance traveled by the master is directly proportional to the elapsed time. Therefore, d_m in Equation (7.3) is analogous to time t in Equation (7.4). In other words, the master distance d_m in ratio following mode is analogous to the move time *TM* in time-based normal motion mode.

Referring to the generic acceleration segment shown in Figure 7.11, we can derive the distance traveled by the slave axis using the area under the curve as [6]

$$d_s = \frac{1}{2} d_m (SR_i + SR_f) \tag{7.5}$$

where SR_i and SR_f are the initial and final speed ratios, respectively.

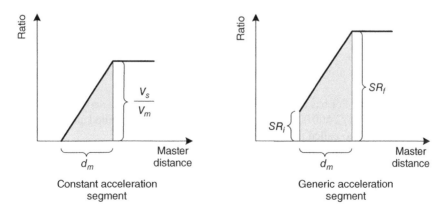

Figure 7.11   Slave axis travel in ratio following

Table 7.1   Distance traveled by the
slave axis in each segment of master
distance in Figure 7.10

Master distance	Slave distance
d_1^a	0
d_2	$d_2/2$
d_3	d_3
d_2	$d_2/2$
d_4	$(d_2 + d_3)^b$

$^a d_1 = L - 2(d_2 + d_3)$, L is the cut
length of the product.
^{b}Return motion.

Going back to the flying knife application, given the velocity profile in Figure 7.10 and Equation (7.5), it is easy to compute the distance that must be traveled by the slave axis in each of the master axis segments (Table 7.1).

Normally, each move in a motion program can be specified by a move time (*TM*) and the distance to be traveled by the axis during that time. Programming in the ratio following mode is similar except the move time is replaced by the master distance (*MD*) before commanding the slave axis position in each segment of the motion profile [5]. Algorithm 7 shows how the motion in Figure 7.10 can be programmed in the ratio following mode. The external master encoder is connected to the motor #1 encoder input on the motion controller. This input is assigned to the *X*-axis and is defined as the master axis. Motor #2 runs the slave axis and is assigned to the *Y*-axis. Every motion segment of the master axis is specified with the *MD* variable followed by the slave motion to be completed in that segment. The knife solenoid, connected to I/O pin 1, is activated as soon as the synchronized segment is reached. The knife is lowered for 300 ms and retracted automatically while the material and the knife are traveling in synchronization.

Time-base Control [12]
In master/slave applications, the slave axis position needs to be programmed as a function of an external master axis *position*. However, all of the computed position trajectories are functions of *time*. Hence, all move commands also describe motion as a function of time. In time-base control, the elapsed time is made proportional to the *distance traveled by the master axis* [12]. This allows the motion program for the slave axis to be still written as if all moves were with respect to time.

Every motion controller has a servo update rate. For example, if the servo update rate is 2.5 kHz, every 400 μs (= 1/2 500 000) the trajectory equations are computed and the servo loops are updated. This is also called the *servo cycle* or sampling time (T_s).

The motion trajectory for an axis is computed using discrete-time equations such as

$$V_{k+1} = V_k + a\Delta t \tag{7.6}$$

where k is the sample number ($k = 0, 1, 2, ...$), "a" is the acceleration, V_k is the velocity at the kth sample, and V_{k+1} is the velocity computed for the next point of the trajectory (($k + 1$)th

---

Algorithm 7 Multiaxis motion: *Flying knife with ratio following* [5]

	Define master axis	
1:	$X \leftarrow EncMotor1$	▷ Motor 1 encoder assigned to X axis
2:	$MasterX \leftarrow X$	▷ Define X-axis as master encoder
	Setup parameters	
3:	$Y \leftarrow Motor2$	▷ Motor 2 assigned to Y (slave) axis
4:	$d1, d2, ..., L \leftarrow Values$	▷ Assign values to master distances
5:	$SRrtn \leftarrow Value$	▷ Set return speed ratio
6:	**procedure** MAIN	
7:	$MoveMode \leftarrow Linear$	▷ Linear moves
8:	$PosMode \leftarrow Inc$	▷ Incremental positions
9:	$Y0$	▷ Move slave axis to beginning point
10:	**while** $(Start = 1)$ **do**	▷ Machine started
11:	$\quad MD \leftarrow d1$	▷ Dwell
12:	$\quad Y(L - (2 * d2 + d3 + d4))$	
13:	$\quad MD \leftarrow d2$	▷ Accel
14:	$\quad Y(d2/2)$	
15:	$\quad KnifeSolenoid \leftarrow 1 : 300$	▷ Activate knife (I/O pin 1) for 300 ms
16:	$\quad MD \leftarrow d3$	▷ Sync
17:	$\quad Y(d3)$	
18:	$\quad MD \leftarrow d2$	▷ Decel
19:	$\quad Y(d2/2)$	
20:	$\quad MD \leftarrow d4$	▷ Return
21:	$\quad Y(-SRrtn * d4/2)$	
22:	**end while**	
23:	**end procedure**	

---

sample). *Time base* is the Δt used in such trajectory calculations. It is the *elapsed time* between the consecutive samples. Equation (7.6) is nothing but the numerical computation of $a = \frac{dV}{dt}$.

The elapsed time does not have to be equal to the real time. Time-base control works by changing the value of the elapsed time Δt. To examine the effect of this, let us look at a simplified example. Assume that $a = 2$, $\Delta t = 1$, and $V_0 = 0$. The controller computes Equation (7.6) at each servo interrupt. For this simplified example, let us also assume that the servo interrupts are every millisecond.

By default, the controller sets the time base equal to the servo interrupt time ($\Delta t = 1$ in this example). Then, we can compute the following velocities for three servo cycles as:

$$V_1 = 0 + 2 \cdot 1 = 2$$
$$V_2 = 2 + 2 \cdot 1 = 4$$
$$V_3 = 4 + 2 \cdot 1 = 6$$

In the first interrupt ($k = 0$), V_1 is calculated as the new velocity. The next calculation occurs 1 ms later ($k = 1$) and V_2 is calculated. At the next interrupt ($k = 2$), the controller updates the velocity to V_3.

Now, let us compute the same three servo cycles but this time using $\Delta t = 0.5$

$$V_1 = 0 + 2 \cdot 0.5 = 1$$
$$V_2 = 1 + 2 \cdot 0.5 = 2$$
$$V_3 = 2 + 2 \cdot 0.5 = 3$$

Even though we are still computing the same equation every servo interrupt at each 1 ms, the final velocity V_3 turns out to be different. The effect is that if the time base is halved, the final velocity is also halved ($V_3 = 3$ as opposed to $V_3 = 6$). Then, we observe that it is possible to control the speed of motion by just changing the time base of its coordinate system.

To establish the relationship between master encoder and time, we must alter the default settings for the elapsed time Δt. If we replace the time source with the master encoder position counts and use the inverse of the master axis maximum speed as the scale factor, we can redefine the elapsed time as

$$\Delta t = \frac{1}{\omega_{master}^{nom}} \cdot \theta_{master} \tag{7.7}$$

where ω_{master}^{nom} is the nominal speed of the master axis when it is running at its expected operating speed (cts/ms). θ_{master} is the distance traveled by the master axis measured by the master encoder (cts). When the master axis is running at nominal speed, the distance traveled in unit time will match the nominal speed. Therefore, Δt will be equal to one. However, if the master axis is running at half of its nominal speed, the distance traveled will be halved. When scaled by the nominal speed as in Equation (7.7), the time-base will also be halved ($\Delta t = 0.5$). As a result, the slave axis moves in its motion program will operate at half the speed, which will maintain both the axes synchronized.

Recall that Algorithm 7 showed how the flying knife application in Figures 7.9 and 7.10 could be programmed using the ratio following method. Now, let us look at how the same application can be programmed using the time-base control method. Algorithm 8 starts by defining the master encoder counts as coming into the X-axis and sets up motor 2 as the following Y-axis. The nip rolls have their own drive and motor that spins at a nominal speed of MS in cts/ms. The main difference in time-base control is that the move commands are specified as a function of time (TM) instead of master distance if the motion program is written in the normal real-time motion mode. In each segment of the motion, the master distance in Figure 7.10 is first converted into a move time using the master speed. For example, the acceleration move time is calculated as $TM = (d2/2)/MS$ in millisecond. Then, the slave axis Y is commanded to move through the slave axis distance in that segment with the specified move time. Again, in the acceleration segment, the slave axis moves through $d2/2$ counts, hence the command $Y(d2/2)$. Even though the program is written as if the motion is happening in real time, the time-base setting (SF) internally makes the slave axis speed adjustable to the master encoder speed.

■ **EXAMPLE 7.4.1 (Spool Winding)**

Spool winding is a common industrial application where strands of wire, textile, fiber, etc., are wound on a rotating spool as shown in Figure 7.12. Usually, the spool has its own drive and motor that spins at a nominal speed of ω_{spool} *rpm*. The traverse axis is controlled by the motion controller and needs to be synchronized to the spool. The master encoder is on the

---

Algorithm 8 Multiaxis motion: *Flying knife with time-base control*

Define master axis and time-base

1:	$X \leftarrow EncMotor1$	▷ Motor 1 encoder assigned to X axis
2:	$MasterX \leftarrow X$	▷ Define X-axis as master encoder
3:	$SF \leftarrow 1/MS$	▷ Time-base scale factor (ms/cts)

Setup parameters

4:	$Y \leftarrow Motor2$	▷ Motor 2 assigned to Y (slave) axis
5:	$d1, d2, ..., L \leftarrow Values$	▷ Assign values to master distances
6:	$SRrtn \leftarrow Value$	▷ Set return speed ratio
7:	**procedure** MAIN	
8:	$MoveMode \leftarrow Linear$	▷ Linear moves
9:	$PosMode \leftarrow Inc$	▷ Incremental positions
10:	$TA \leftarrow 10$	▷ 10 ms accel time
11:	**while** $(Start = 1)$ **do**	▷ Machine started
12:	$Dwell \leftarrow (L - (2 * d2 + d3 + d4))/MS$	▷ Dwell (ms)
13:	$TM \leftarrow (d2/2)/MS$	▷ Accel
14:	$Y(d2/2)$	
15:	$KnifeSolenoid \leftarrow 1 : 300$	▷ Activate knife (I/O pin 1) for 300 ms
16:	$TM \leftarrow d3/MS$	▷ Sync
17:	$Y(d3)$	
18:	$TM \leftarrow (d2/2)/MS$	▷ Decel
19:	$Y(d2/2)$	
20:	$TM \leftarrow d4/MS$	▷ Return
21:	$Y(-SRrtn * d4/2)$	
22:	**end while**	
23:	**end procedure**	

---

spool motor but it is also connected to the motor #1 encoder of the motion controller. This is how the motion controller receives the master position counts (θ_{master} in Equation (7.7)). The traverse axis also has an encoder measuring its position in cts/mm of linear distance. The pitch of the winding is P, which corresponds to the linear distance between adjacent strands in one revolution of the spool. The width of the spool is W. Each layer consists of laying the material on the spool once back-and-forth along the width of the spool.

Solution

Algorithm 9 shows spool winding with time-base control [2]. We start by routing the master encoder (spool encoder) data into the X-axis in the software and declaring it as the master axis. MS is the spool nominal speed ω_{master}^{nom} in cts/ms in Equation (7.7). The scale factor SF is the inverse of the master axis nominal speed in ms/cts. The Y-axis is set up as the traverse axis. The *TimePerLayer* is how long it takes to wind one layer (back-and-forth) of material on the spool. The controller commands the Y-axis to make a move through the width of the spool in TM milliseconds. Then, it commands the Y-axis to return through the same distance ($-Width$), which completes one layer of material on the spool. The motion is repeated by the WHILE loop as long as the machine is running.

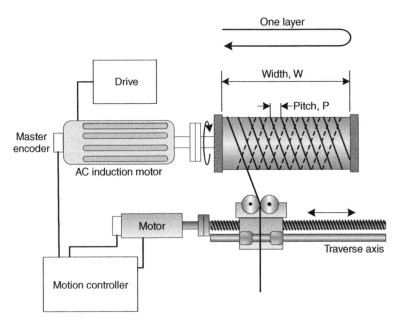

Figure 7.12   Spool winding to wind strands on a spool

Algorithm 9 Multiaxis motion: *Spool winding with time-base control* [2]
___
Define master axis and time-base
1: $X \leftarrow EncMotor1$ ▷ Motor 1 encoder assigned to X axis
2: $MasterX \leftarrow X$ ▷ Define X-axis as master encoder
3: $SF \leftarrow 1/MS$ ▷ Time-base scale factor (ms/cts)
    Setup parameters
4: $Y \leftarrow Motor2$ ▷ Motor 2 assigned to Y (traverse) axis
5: $P \leftarrow pitch$ ▷ Motion parameters
6: $W \leftarrow width$
7: $MS \leftarrow MasterSpeed$ ▷ Spool speed in rev/s
8: **procedure** MAIN
9:      $MoveMode \leftarrow Linear$ ▷ Linear moves
10:     $PosMode \leftarrow Inc$ ▷ Incremental positions
11:     $TM \leftarrow W/(P * MS)$ ▷ Move time for Y axis
12:     **while** $(Start = 1)$ **do**   ▷ Machine started
13:         $Y(W)$ ▷ Begin winding a layer
14:         $Y(-W)$ ▷ Return winding to finish the layer
15:     **end while**
16: **end procedure**
___

■ EXAMPLE 7.4.2 Rotating Knife

Continuous material, such as paper, plastics, wood, etc., can be cut to length using a rotating knife as shown in Figure 7.13.

Typically, the web is not under the control of the motion controller but the knife is. Once the machine is started, the web moves at a constant speed. There are two phases of the knife motion: (1) Cutting, and (2) Adjustment. In the cutting phase, the knife is in contact with the material. The tangential speed of the rotary knife must be synchronized to the linear speed of the web. In the adjustment phase, the knife is not in contact with the material. It rotates back to the beginning of the cutting position while the web continues to advance. The cut length of the material can be adjusted by speeding up or slowing down the knife in the adjustment phase.

Write a motion control program to synchronize the knife to the web, which moves at V_{web} linear speed. The desired cut length is L_d and the material moves L_c length in the cutting phase. The radial distance measured from the center to the knife edge is r_k. The desired cut length is shorter than the circumference of the rotating knife.

Solution

If the knife rotates at a constant speed, the cut length will be equal to the circumference of the knife. If the desired cut length is shorter than the circumference of the rotating knife, the knife must run faster (advance) in the adjustment phase as shown in Figure 7.14. However, if the desired cut length is longer than the circumference of the knife, it must run slower (retard) in the adjustment phase. The programmable master/slave synchronization capability of the motion controller enables such variable length cuts. If the knife and the web were coupled mechanically (using pulley-and-belt or gearing), the cut length would be fixed.

The cycle time to cut the desired length L_d $(= L_c + L_a)$ can be found from

$$t_{cyc} = \frac{L_d}{V_{web}}$$

Similarly, the cutting time is

$$t_c = \frac{L_c}{V_{web}}$$

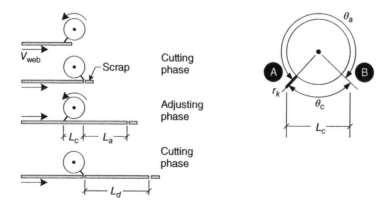

Figure 7.13   Rotating knife to cut continuous material (web)

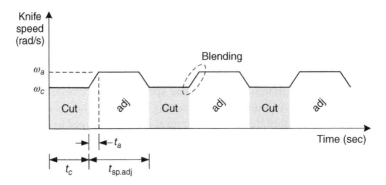

Figure 7.14 Rotating knife motion profile to cut material lengths that are shorter than the knife circumference

Therefore, the time left to adjust the knife speed is

$$t_{\text{sp.adj}} = t_{\text{cyc}} - t_c$$

The rotational speed of the knife in the cutting phase is

$$\omega_c = \frac{\theta_c}{t_c}$$

$$= \frac{L_c / r_k}{t_c}$$

Similarly, rotational speed of the knife during the adjustment phase is

$$\omega_a = \frac{\theta_a}{t_{\text{sp.adj}} - t_a}$$

$$= \frac{2\pi - \theta_c}{t_{\text{sp.adj}} - t_a}$$

Algorithm 10 shows the motion program for this application, which uses the time-base control approach. The web speed is measured by a master encoder. This is typically done either by using the encoder on the nip rolls or using an encoder wheel on the web. The master encoder data is routed into the X-axis in the software and declared as the master axis. *MS* is the nominal web speed $\omega_{\text{master}}^{\text{nom}}$ in cts/ms in Equation (7.7). The scale factor *SF* is the inverse of the master axis nominal speed in ms/cts. The knife encoder data comes into the A-axis, which is a rotary axis in the software. Assuming the knife is at position "A", the controller commands the B-axis to make a move from position "A" to "B" to make the cut in t_c milliseconds. Then, it advances the knife counterclockwise (CCW) from position "B" to "A" in $t_{\text{sp.adj}}$ milliseconds. The cycle repeats as long as the machine is enabled.

Depending on how a particular controller does blending, the t_c and $t_{\text{sp.adj}}$ times may need to be adjusted to incorporate the acceleration/deceleration times between the speed changes. Also, the acceleration time t_a is kept to a minimum for quick acceleration to the new knife speed in the adjustment phase.

Algorithm 10 Multiaxis motion: *Rotating knife with time-base control*

    Define master axis and time-base

1: $X \leftarrow EncMotor1$ ▷ Motor 1 encoder assigned to X axis

2: $MasterX \leftarrow X$ ▷ Define X-axis as master encoder

3: $SF \leftarrow 1/MS$ ▷ Time-base scale factor (ms/cts)

    Setup parameters

4: $A \leftarrow Motor2$ ▷ Motor 2 assigned to A rotary axis (knife)

5: $PosA \leftarrow thetaA$ ▷ Angular position of point A (cts)

6: $PosB \leftarrow thetaB$ ▷ Angular position of point B (cts)

7: $Tcut \leftarrow tc$ ▷ Cutting time (ms)

8: $Tadj \leftarrow tsp.adj$ ▷ Speed adjustment time (ms)

9: $MasterSpeed \leftarrow MS$ ▷ Web speed in/ms

10: **procedure** MAIN

11:     $MoveMode \leftarrow Linear$ ▷ Linear moves

12:     $PosMode \leftarrow Abs$ ▷ Absolute positions

13:     $TA \leftarrow 0.1 * Tadj$ ▷ Accel time set to 10% of adjustment time (ms)

14:     **while** $(Start = 1)$ **do**   ▷ Machine started

15:         $TM \leftarrow Tcut$ ▷ Move time set to cutting time

16:         $A(PosB)$ ▷ Move axis-A from pos. A to B (CCW) to make a cut

17:         $TM \leftarrow Tadj$ ▷ Move time set to adjustment time

18:         $A(PosA)$ ▷ Move axis-A from pos. B to A (CCW) to advance the knife

19:     **end while**

20: **end procedure**

7.4.3.3   Triggered Camming

The electronic camming techniques presented so far enable synchronizing the motion of one or more axes to the motion of a master axis. Speeds of the axes are matched as in the case of the flying knife application with moving web material such as blank roll of paper. However, in some applications, not only the speeds but also the alignment of one axis with respect to the other must match. Consider a continuous roll of paper with a preprinted logo every 18 in. The paper must be cut between the logos so that each cut piece contains a preprinted logo. Typically, a registration mark indicates the cutting line between the printed segments. Hence, simply matching the speed of the flying knife to the master will not be sufficient as this will not guarantee cutting the material at the registration marks. The alignment of one axis with respect to a point on the other is called *phase adjustment* [14, 15]. A phase shift can be commanded while the axes are in synchronization ($SR = 1$). This can be in the form of advancing or retarding the position of the slave axis. The shift motion is superimposed on the already existing synchronized motion between the axes.

    In triggered camming applications, a sensor is used to detect a registration mark. When the mark is detected, a high-speed position capture input records the instantaneous position of the master axis. From this accurately measured position reference, a preset master axis travel distance is programmed as a position trigger point. As the registration mark moves and reaches this trigger point, the controller begins to accelerate the slave axis so that in a short while it is perfectly synchronized and aligned with the master. The high-speed position capture is a

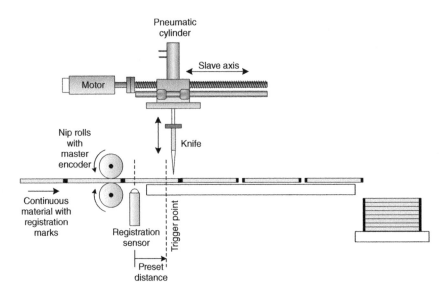

Figure 7.15 Flying knife with registration to cut continuous material at registration marks into fixed lengths

special axis-level I/O line which is able to capture the master encoder count within 1 μs of receiving the sensor input.

Flying knife with registration is one common example of triggered camming. Figure 7.15 shows a modified version of the system in Figure 7.9 with the added registration marks on the web material and the registration mark sensor.

In programming triggered camming, the trigger is first disabled so that the slave motion program can be loaded to the buffer and all moves can be precalculated. Then, the program waits for a trigger while the slave axis is stationary at a starting point. When the trigger comes, the slave axis starts its motion cycle. While the slave axis is moving, often the next set of moves is loaded to the buffer so that the moves can be precalculated. When the slave axis completes its motion, it comes back to the starting point. The next moves have already been calculated. The machine waits for the next trigger to repeat the cycle.

Algorithm 11 shows flying knife with registration. It is a modified version of the ratio following method in Algorithm 7. First, we start with temporarily disabling the trigger point. Then, the moves are loaded to the buffer for the slave axis and computed. The program waits until a trigger is received. The trigger is captured in a *latch event handler* that is running in the background as the controller is multitasking. When the registration sensor detects a mark, the latch event handler is called. It captures the master encoder counts, adds the preset master distance to it and enables the trigger point. The precalculated slave motion is executed as soon as the master position reaches the calculated trigger point. This way, the slave motion is synchronized to a point on the master axis. Meanwhile, the program execution drops out of the WAIT command, goes back to the top of the WHILE loop, disables the trigger, precomputes the next move, and waits for the next trigger.

Algorithm 11 Multiaxis motion: *Flying knife with registration*

Define master axis

1: *X ← EncMotor*1   ▷ Motor 1 encoder assigned to *X* axis

2: *MasterX ← X*   ▷ Define *X*-axis as master encoder

Setup parameters

3: *Y ← Motor*2   ▷ Motor 2 assigned to *Y* (follower) axis

4: *d*1, *d*2, ... *← Values*   ▷ Assign values to master distances

5: *SRrtn ← Value*   ▷ Set return speed ratio

6: *TrigPointDis ← Value*   ▷ Preset distance to trigger point

7: **procedure** MAIN

8:    *MoveMode ← Linear*   ▷ Linear moves

9:    *PosMode ← Inc*   ▷ Incremental positions

10:   *Y*0   ▷ Beginning point

11:   **while** (*Start* = 1) **do**   ▷ Machine started

12:       *Trigger ← 0*   ▷ Trigger disabled

13:       *MD ← d*1   ▷ Dwell

14:       *Y*(*L* − (2 * *d*2 + *d*3 + *d*4))

15:       *MD ← d*2   ▷ Accel

16:       *Y*(*d*2/2)

17:       *KnifeSolenoid ←* 1 : 300   ▷ Activate knife (I/O pin 1) for 300 ms

18:       *MD ← d*3   ▷ Sync

19:       *Y*(*d*3)

20:       *MD ← d*2   ▷ Decel

21:       *Y*(*d*2/2)

22:       *MD ← d*4   ▷ Return

23:       *Y*(−*SRrtn* * *d*4/2)

24:       *WAIT*(*Trigger* = 1)   ▷ Wait until moves are triggered

25:   **end while**

26: **end procedure**

27: **function** LATCHEVENT(*MasterPos*)

Compute trigger point

28:   *TriggerPoint ← (MasterPos + TrigPointDis)*

29:   *Trigger ← 1*   ▷ Enable trigger

30: **end function**

7.4.4 Tension Control

Web handling is a typical application of tension control where a thin, long, flexible material called web needs to be guided through a processing machine. The web can be film, shrink wrap, aluminum foil, currency, paper, carpet, tape, etc. Paper industry, which produces several hundred tons of paper a day just in the United States, is probably the oldest among the web handling industries. Printing industry is the largest web handling industry with products such as newpapers and magazines. Another web handling industry is the converting industry where a supply of raw web material is altered into other products such as embossed rolls of paper towels with perforations.

Web handling requires routing the material through a series of idler and drive rolls to guide it through a processing machine. If the web slips over the rolls, it cannot be steered accurately. Therefore, the material must be tensioned properly to keep it in traction with the rolls. Minor speed differences between the nip rolls can develop slack or extra tension in the material, which may eventually break the web. Proper tension must be applied so that the material does not overstretch and get damaged. Furthermore, if the tension is not controlled, telescoping, coned rolls, and wrinkles can develop in the web.

As shown in Figure 7.16, converting processes typically include three tension zones: (1) Unwind, (2) Internal, and (3) Rewind. Each of these zones is controlled independently and may have its own tension requirements. As the unwind and rewind rolls are processed, their diameters change. To keep the web tension constant, the torque and speed on each roll must be continuously adjusted. During the processing of the web in the internal zone, the tension and speed are fairly constant since the roll diameters in this zone do not change.

Figure 7.17 shows a simplified converting machine where nip rolls, driven by a motor, pull the material from a supply spool to feed the web into the machine at a constant speed. A core is mounted on a motor-driven mandrel that winds the web material on the core. This arrangement is known as *center winding*.

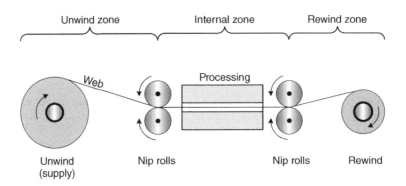

Figure 7.16 Typical converting process with unwind, internal, and rewind tension zones

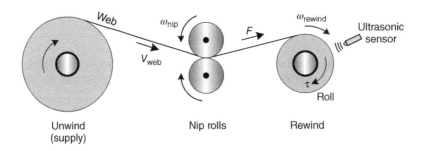

Figure 7.17 Simplified converting machine

To avoid damaging the material, the machine needs to keep the web speed and tension constant. Torque applied by the rewind drive (mandrel) is given by

$$\tau = F r_{roll} \tag{7.8}$$

where F is the web tension and r_{roll} is the radius of the rewind roll. As the material is wound, the roll radius increases. Therefore, to keep the web tension constant, the torque applied by the rewind drive must increase proportionally to the roll radius. Similarly, the braking torque applied by the unwind drive must decrease proportionally to the radius of the supply roll.

The web speed, V_{web}, can be found from the tangential speed of the nip rolls. Since the web speed must be constant, the same tangential speed must be maintained by the winder. We can then write

$$V_{web} = \omega_{nip} r_{nip} = \omega_{rewind} r_{roll} \tag{7.9}$$

where r_{nip} is the radius of a nip roll, ω_{nip} and ω_{rewind} are the angular speeds of the nip rolls and the rewind drive, respectively. We can rewrite this equation as

$$\omega_{rewind} = \omega_{nip} \frac{r_{nip}}{r_{roll}} \tag{7.10}$$

If the nip rolls are operating at a constant speed, then the rewind drive must slow down as the roll radius increases.

7.4.4.1   Open-Loop Web Tension Control

In this approach, the nip rolls are commanded to run at a fixed speed, ω_{nip}, while the rewind axis is controlled in torque mode, which applies a commanded torque to the rewind. The speed of the nip rolls and the rewind axis can be measured from their motor encoders. The radius of the nip rolls is known. If we can measure the roll radius, r_{roll} (e.g., with an ultrasonic sensor), then the torque to be commanded can be calculated from Equation (7.8). If the roll radius cannot be measured, then it can be calculated from Equation (7.9) since the speeds for the nip rolls and the rewind axis will be available from their encoders. Then, the commanded torque can be found from

$$\tau_{cmd} = F r_{nip} \frac{\omega_{nip}}{\omega_{rewind}}$$

where F is the desired web tension set by the user. This approach is considered open-loop since the web tension is not directly measured.

In practice, a different implementation may be used to achieve a similar result. Instead of controlling the rewind axis in torque mode, we can operate it in velocity mode with a torque limit. Since the speed of the nip rolls is known and assuming the radius of the roll is measured, we can calculate the speed that needs to be commanded to the rewind axis from Equation (7.10). To apply tension, a slightly higher speed is commanded as

$$\omega_{cmd} = \omega_{nip} \frac{r_{nip}}{r_{roll}} + \omega_{offset} \tag{7.11}$$

where ω_{offset} is the additional speed adjustable by the user. As the rewind axis tries to reach this slightly higher speed, it applies tension on the web material. A torque limit is set on the axis to prevent the controller from applying excessive torque while trying to reach the commanded speed. Furthermore, if the web breaks, the torque limit keeps the axis from accelerating to dangerously high speeds.

7.4.4.2   Closed-Loop Web Tension Control

The open-loop approach works well if the web is accelerated or decelerated slowly. In applications where high speed operation and better tension control are necessary, closed-loop tension control is employed. The controller is configured in *cascaded loops* configuration where the output of an outer tension control loop becomes the input command of an inner velocity loop. To regulate the tension in closed-loop control, we must measure it. There are two common measurement implementations in practice: (1) Load cell and (2) Dancer roll.

In the *load cell* implementation, the controller tries to maintain a commanded tension by comparing it to the actual web tension measured by the load cell (Figure 7.18). The tension error is used as a speed command to the winder axis.

In the *dancer roll* approach, the web is in contact with a dancer roll, which applies tension on the web. In older designs, the tension was adjusted by adding or removing weights attached to the dancer roll. In newer machines, pneumatic cylinders are used to apply pressure on the dancer roll [11]. The air pressure in the cylinder, hence the desired tension on the web, can be set very accurately by using electronic pressure regulators with built-in closed-loop pressure sensing/regulating capabilities. When the dancer is at its neutral position, the dancer force is balanced by the web tension. If the web tension matches the dancer force, the dancer remains at the neutral position. If the web tension changes, the dancer moves up or down. The *position* of the dancer roll is measured with an encoder attached to it.

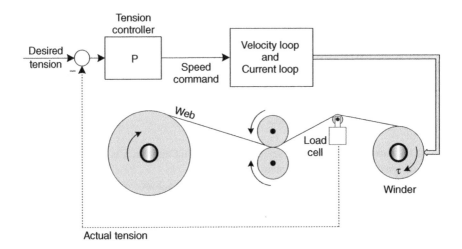

Figure 7.18   Closed-loop web tension control with load cell

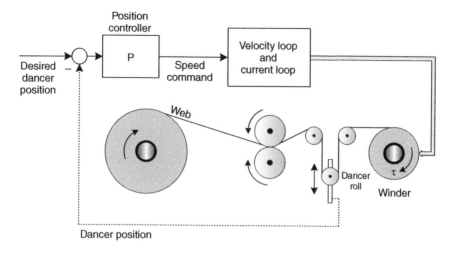

Figure 7.19 Closed-loop web tension control with dancer roll

The desired web tension is set by the dancer roll weight or force but it is maintained by a position loop as shown in Figure 7.19. The dancer position error, which corresponds to tension error, is then used to regulate the commanded speed to the rewind axis. If the dancer is moving up, then the tension is increasing and the rewind axis is slowed down. If the dancer is moving down, then the tension is decreasing hence the rewind axis is sped up.

The dancer roll approach has a damping effect on the tension control to smooth out fluctuations. Therefore, it can accommodate variations in the web material or jumps between material layers as in winding cables or thick wires. Furthermore, it can act as a storage area to accumulate some extra length of material while the web is running through the machine.

7.4.4.3 Taper Tension

As the radius of the rewind roll increases, each additional layer builds up compressive forces on the inner layers. As a result, the layers near the core can wrinkle or the core can get crushed. Taper tension is a way to address this problem by decreasing the web tension as the radius of the roll increases. Taper tension is used only in the rewind zone.

Figure 7.20 shows constant tension, taper tension, and the corresponding rewind torque in each case. As mentioned before, to keep the web tension constant, the torque must be increased *linearly* as the radius of the rewind roll increases (Equation (7.8)).

The taper tension linearly decreases from an initial tension of F_i at the core radius r_{core} to F_f at the final roll diameter r_{roll}. At any radius r, the torque for taper tension is given by

$$\tau_{cmd} = \left[F_i + \left(\frac{F_f - F_i}{r_{roll} - r_{core}} \right) (r - r_{core}) \right] \cdot r \tag{7.12}$$

In taper tension, the torque must still be increased as the roll radius increases. However, the required torque profile is *nonlinear* as shown in Figure 7.20.

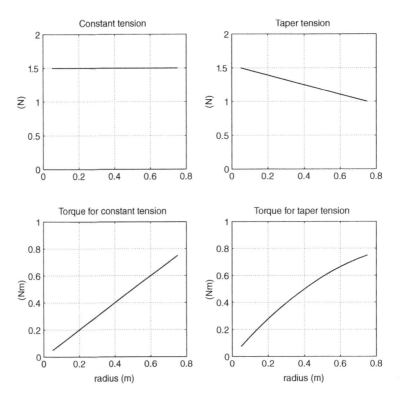

Figure 7.20   Constant tension, taper tension, and the corresponding rewind torques ($F_i = 1.5\,\text{N}, F_f = 1\,\text{N}, r_{core} = 0.05\,\text{m}, r_{roll} = 0.75\,\text{m}$)

7.4.5   Kinematics

Machines built for positioning systems, such as pick-and-place operations, have a tool they position in 2D or 3D space. In such machines, we have *tool tip coordinates* and *joint coordinates*. Consider the gantry machine in Figure 7.6a. The tool tip coordinates are the X, Y, Z coordinates of the tip of the tool that interacts with the environment. However, the rotational position of each motor shaft is the joint coordinate of that axis. The tool tip coordinates can be in *inches* whereas the joint coordinates are often in motor counts (cts).

If each motor is rotated through a certain number of counts where will be the tool tip and what will be its orientation? This is known as the *forward kinematics* problem. Specifically, computation of the tool tip position and orientation from a given a set of joint coordinates is the forward kinematics [10].

Inverse kinematics problem requires finding joint coordinates that would result in positioning the tool tip at a desired location in a desired orientation in 3D space. As the name implies, this problem is the opposite of the forward kinematics problem. It is usually much more complicated than the forward kinematics problem.

In a positioning system, it is much more natural for the user to specify the motion in the X, Y, Z tool tip coordinates as the tool goes from point A to point B. However, the motion controller works in the joint coordinates since it commands the motors in counts. This requires a capability to compute the inverse kinematics for the machine.

Most machines for positioning are built using *Cartesian mechanisms* with three perpendicular axes. In these mechanisms, one axis is aligned with the X direction, another with the Y direction, and the third with the Z direction of a 3D Cartesian coordinate frame. The desired tool tip position is then specified as a point in this Cartesian coordinate frame. This makes the inverse kinematics problem simple since in Cartesian mechanisms the inverse kinematics is either a simple scale factor or a linear equation. In the gantry machine in Figure 7.6a, if the ball-screw of the base axis advances $X = 1$ inch in 8000 cts of its motor rotation, the inverse kinematics can be computed from $8000X$. If we want the axis to move 2 in, the motor must be commanded 16 000 cts. Similarly, to move the cross axis $Y = 3$ in, the motor for the Y-axis would be commanded 24 000 cts. The inverse kinematics is simple because it maps each coordinate of the tool tip to a separate motor through a simple scale factor. This makes Cartesian mechanisms very popular. Furthermore, Cartesian mechanisms do not generate any change in the inertia while in motion, which makes their design and motor sizing simpler.

With recent advances in the motion controllers and more computing power, *non-Cartesian mechanisms* are made possible. Most common ones are robotic mechanisms including the SCARA and delta robots shown in Figure 7.21. These mechanisms can provide significant improvements in the production rates over Cartesian mechanisms for the same application. They tend to have a much larger workspace compared to the footprint of a Cartesian mechanism for the same work space. Furthermore, in Cartesian mechanisms at least one axis must carry another which requires bigger motors and more power.

In a non-Cartesian, mechanism the inverse kinematics involves complex, nonlinear set of equations that have to be solved in real time. In fact, sometimes it is not possible to solve these equations in a closed form. Furthermore, multiple solutions may exist or a solution may not exist at all.

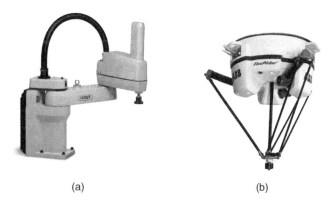

(a) (b)

Figure 7.21 SCARA and delta robots are examples of non-Cartesian mechanisms. (a) SCARA robot. Copyright © 2015, Adept Technology, Inc. (Reproduced by permission of Adept Technology, Inc.) [8]. (b) Delta robot (Reproduced by permission of ABB Corp.) [7]

Motion controllers handle inverse kinematics calculations for non-Cartesian mechanisms in one of the following ways:

1. *Compute on a host computer*: Due to the complexity of the equations, they may be computed by a program running on a host computer connected to the motion controller. This program converts the tool tip coordinates into motor positions and sends them to the motion controller through a communication link. The controller then simply executes the desired motor moves.

2. *Compute in the motion program*: Typically, kinematic updates are required every 5 to 10 ms. If the motion controller has a fast processor, it may be possible to incorporate the inverse kinematics equations directly into the motion program. In this approach, first a set of tool tip positions along the desired tool trajectory are generated by the program. Then, these positions are used to compute the corresponding joint positions to populate a look-up table. Next, adjacent joint positions are used in pairs to calculate a path between them using contour commands. This step converts the absolute joint positions in the look-up table into incremental contour data and defines a smooth path that connects all joint positions into a joint trajectory for each motor. Finally, these trajectories are used to make contour moves by each joint. The result is the tool tip following the desired path in 3D space [3].

3. *Emulate Cartesian mechanism*: In this approach, the motion controller computes the forward and inverse kinematics equations that are placed into subroutines in special buffers [12]. The main motion program is written as if the mechanism were a Cartesian mechanism. This makes the motion programming very straightforward. For example, to move the tool tip 2 inch in the X direction and 3 inch in the Y direction, we simply enter $X2\ Y3$ in the main program. When the motion program is first executed, the controller automatically calls the forward kinematics subroutine to compute the current tool tip location, given the joint positions at the instant the program was started. This establishes the starting coordinates of the tool tip. Then, as it continues to execute the motion program, it calls the inverse kinematics subroutine when it encounters a line such as $X2\ Y3$ in the program. The corresponding joint moves are computed and commanded to the motors.

4. *Use plug-in libraries*: Some motion controllers have inverse kinematics routines developed for specific types of common robotic mechanisms. Vendors supply these as plug-in library modules [16, 20]. The motion controller software can call the module or function corresponding to the mechanism it is controlling to compute the inverse kinematics.

5. *Embed into firmware*: The forward and inverse kinematics equations for a particular machine can be programmed directly into the firmware in the motion controller. This is a low-level embedded software that runs at a very high speed allowing kinematic updates at very high rates (such as at 32 kHz). The user can then write the motion program in tool tip coordinates. Since this approach customizes the motion controller for a very specific type of machine, it requires working closely with the motion controller vendor [13].

7.4.5.1   SCARA Robot

As mentioned earlier, one of the most common non-Cartesian mechanisms used in positioning applications is the SCARA robot arm. As shown in Figure 7.22a, this robot has four axes. Axes 1, 2, and 3 are rotational. Axis 4 is translational. The tool tip (fingertips of the gripper) can be

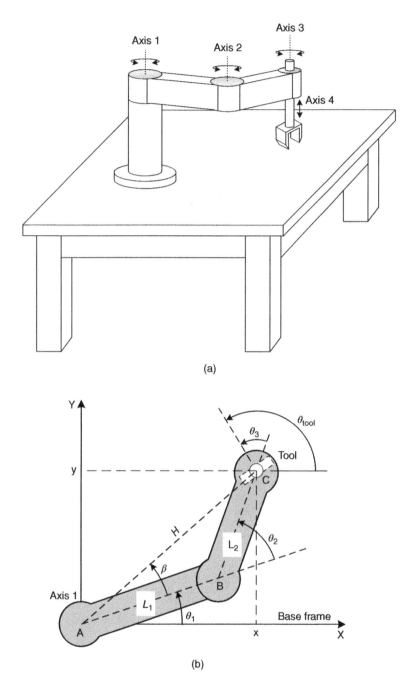

(a)

(b)

Figure 7.22   Four-axis SCARA robot for kinematic analysis. (a) Four-axis SCARA robot. (b) Top view

positioned at a point with x, y, z coordinates with respect to a Cartesian frame attached to the base of the robot. Furthermore, angle θ_{tool} with respect to the base frame is specified by the user to designate the desired tool orientation (Figure 7.22b).

Refering to Figure 7.22b, the forward kinematics equations for this robot can be written as [10]

$$
\begin{aligned}
x &= L_1 \cos\theta_1 + L_2 \cos(\theta_1 + \theta_2) \\
y &= L_1 \sin\theta_1 + L_2 \sin(\theta_1 + \theta_2) \\
z &= z_0 + d_4 \\
\theta_{tool} &= \theta_1 + \theta_2 + \theta_3
\end{aligned}
\tag{7.13}
$$

where θ_1, θ_2, and θ_3 are known joint positions measured by encoders. L_1 and L_2 are the link lengths measured between joint axes along the common normal between them. The linear displacement of axis 4 is measured by d_4. When this axis is at its home position, the tool tip is at z_0 distance from the horizontal plane of the base frame.

The inverse kinematics solution requires finding $\theta_1, \theta_2, \theta_3$, and d_4 given the desired x, y, z coordinates of the tool tip and the tool orientation θ_{tool} with respect to the base frame. Using the ABC triangle and the law of cosines, we can write

$$
H^2 = L_1^2 + L_2^2 - 2L_1 L_2 \cos(180 - \theta_2)
$$

Note that $\cos(180 - \theta_2) = -\cos(\theta_2)$. Then,

$$
\cos(\theta_2) = \frac{H^2 - L_1^2 - L_2^2}{2L_1 L_2}
\tag{7.14}
$$

where $H = \sqrt{x^2 + y^2}$. Similarly, the law of cosines can be used to find angle β

$$
\cos(\beta) = \frac{H^2 + L_1^2 - L_2^2}{2L_1 H}
\tag{7.15}
$$

From the x, y coordinates of the tool tip, we can find

$$
\tan(\theta_1 + \beta) = \frac{y}{x}
\tag{7.16}
$$

Using Equations (7.14), (7.15), and (7.16), we can find the set of inverse kinematics equations as

$$
\begin{aligned}
\beta &= \cos^{-1}\left(\frac{H^2 + L_1^2 - L_2^2}{2L_1 H}\right) \\
\theta_1 &= \tan^{-1}\left(\frac{y}{x}\right) - \beta \\
\theta_2 &= \cos^{-1}\left(\frac{H^2 - L_1^2 - L_2^2}{2L_1 L_2}\right) \\
\theta_3 &= \theta_{tool} - \theta_1 - \theta_2 \\
d_4 &= z - z_0
\end{aligned}
\tag{7.17}
$$

Algorithm 12 shows the motion program for a SCARA robot. The algorithm uses the emulation of Cartesian mechanism approach mentioned earlier. The main motion program simply defines a motion profile and commands to move to the Cartesian positions (700, 0) and (400, 300) in millimeters as if the mechanism were a Cartesian mechanism. The *FwdKin* subroutine implements Equation (7.13). Similarly, the *InvKin* subroutine contains the inverse kinematics equations from (7.17). In line 27 of the algorithm, *ATAN2* function is used. This is a built-in inverse tangent function in most controllers to compute the sign of the angle correctly by taking

---

Algorithm 12 Multiaxis motion: *SCARA robot* [1, 12]

1: **procedure** MAIN
   Setup parameters
2:   $L1, L2, L3, z0, \leftarrow \dots$ ▷ Define all geometric dimensions of the robot
3:   $C1 = L1^2 + L2^2$ ▷ Calculate the constant terms
4:   $C2 = 2 * L1 * L2$
5:   $C3 = L1^2 - L2^2$
6:   $I \leftarrow Motor1$ ▷ Motor 1 assigned to inv. kinematics
7:   $I \leftarrow Motor2$ ▷ Motor 2 assigned to inv. kinematics
8:   $I \leftarrow Motor3$ ▷ Motor 3 assigned to inv. kinematics
9:   $I \leftarrow Motor4$ ▷ Motor 4 assigned to inv. kinematics
10:   $TA \leftarrow 100$ ▷ 100 ms accel time
11:   $TM \leftarrow 900$ ▷ 900 ms move time
12:   $MoveMode \leftarrow Linear$ ▷ Linear moves
13:   $PosMode \leftarrow Abs$ ▷ Absolute positions
14:   $X700\ Y0$ ▷ Move to 700, 0
15:   $Dwell \leftarrow 20$ ▷ Dwell 20 ms
16:   $X400\ Y300$ ▷ Move to 400, 300
17: **end procedure**

18: **function** FWDKIN$(T1, T2, T3, d4)$
   Compute forward kinematics
19:   $X = L1 * COS(T1) + L2 * COS(T1 + T2)$
20:   $Y = L1 * SIN(T1) + L2 * SIN(T1 + T2)$
21:   $Z = z0 + d4$
22:   $Ttool = T1 + T2 + T3$
23: **end function**

24: **function** INVKIN$(x, y, z, Ttool)$
   Compute inverse kinematics
25:   $Hsquare = x^2 + y^2$
26:   $Beta = ACOS(Hsquare + C3)/(2 * L1 * SQRT(Hsquare)))$
27:   $T1 = ATAN2(y, x) - Beta$
28:   $T2 = ACOS((Hsquare - C1)/C2)$
29:   $T3 = Ttool - T1 - T2$
30:   $d4 = z - z0$
31: **end function**

---

the angle quadrant into account. At the beginning of the program, the *FwdKin* subroutine is called once to find the initial tool tip position in Cartesian coordinates with the current motor positions. Then, the *InvKin* subroutine is called repeatedly at each servo cycle to determine the motor position updates as the tool tip is incrementally advanced along the path generated by a move command such as in line 14 in the main program.

Problems

1. On a machine control panel, there is a X-FORWARD push button and a FWD pilot light. The button is used to jog the X-axis of the machine in the forward direction. Modify the PLC program in Algorithm 3 so that when the X-FORWARD button is pressed, the FWD light turns on and the X-axis jogs in the forward (positive) direction. The light remains on as long as the button is held pressed. When the button is released, the axis stops and the light turns off. In each case, the commands should be issued only once (edge-triggered).

   Use JOG+ command to jog the axis. The JOG/ command stops the axis. The X-FORWARD button is connected to input $INP1$ and the FWD light is connected to output $OUT2$. The push button is normally-open, momentary-contact type. When the button is pressed, the $INP1$ input goes high. Motor #1 is assigned to the X-axis.

2. On a machine control panel, there is a RUN button and a STOP button. The buttons are used to jog or stop the X-axis of the machine. Modify the PLC program in Algorithm 3 so that when the STOP button is not pressed and the RUN button is pressed, the X-axis starts jogging. While the axis is in motion, if the STOP button is pressed, the axis stops. The JOG commands should be issued only once (edge-triggered).

   Use JOG+ or JOG/ commands to run or stop the axis. The RUN button is connected to input $INP1$ and the STOP button is connected to input $INP2$. The RUN button is normally-open, momentary-contact type. When the RUN button is pressed, the $INP1$ input goes high. The STOP button is normally-closed, maintained-contact type. When the STOP button is pressed, the $INP2$ input goes low. Motor #1 is assigned to the X-axis.

3. There is a READY light on a machine control panel. When the machine finishes its current task, the READY light is to blink indicating that the machine is waiting for its next command from the user. As the light is blinking, it is on for 1 s and off for 1 s. Write a PLC program as in Algorithm 3 and use a timer to blink the light. The READY light is connected to output $OUT10$. If $OUT10$ is high ($OUT10 = 1$), the light turns on. The controller servo cycle is 5 ms.

4. A SCARA wafer handling robot used in semiconductor industry is shown in Figure 7.23. The robot has two rotational axes (θ_1 and θ_2) and one linear vertical Z-axis. Link lengths are $L_1 = 12$ in and $L_2 = 10$ in. The robot picks up a processed silicon wafer from the machine at point P and puts it into a wafer inspection tool at point T. It picks up the wafer by vacuum.

   Write a motion program similar to Algorithm 12 to transfer a wafer from point P to T when a TRANSFER push button on the control panel is pressed. The TRANSFER button is connected to the $INP1$ input of the motion controller. The vacuum tool is activated by setting the controller output $OUT10 = 1$. The robot starts from and goes back to its home position in each cycle. The tool tip is considered to be at point C. Hence, when the robot moves to P, points C and P coincide.

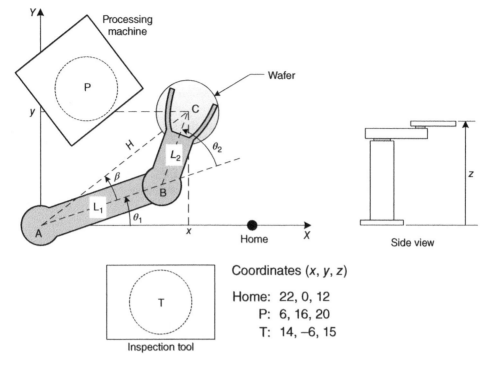

Coordinates (x, y, z)

Home: 22, 0, 12
    P: 6, 16, 20
    T: 14, –6, 15

Figure 7.23 SCARA silicon wafer handling robot for Problem 4

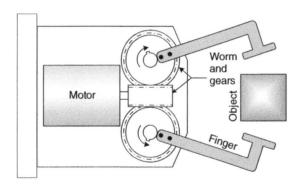

Figure 7.24 Force-controlled robot gripper for Problem 5

5. A force-controlled robot gripper shown in Figure 7.24 is under development. The gripper motor is connected to the *Motor*1 channel of a motion controller. When the motor shaft rotates, the worm-gear arrangement makes the fingers open or close. The design goal is to apply a certain level of grasping force set by the software.

   The control panel has a GRASP push button connected to the input variable *INP*1. When the button is pressed and held, the fingers close. When it is released, the fingers open. The fingers rotate $30°$, which corresponds to 15 000 cts from the motor encoder. The motor speed that results in the desired finger speed is 30 000 cts/s. The motor torque that leads to the desired finger grasp force is 1 lb in. To control the applied force, set a torque limit for the motor and operate it in velocity mode (jog it) with a slightly higher speed as explained in Section 7.4.4.1. $JOG + Motor1$ can be used as the command to jog the motor to close the fingers. Similarly, $JOG - Motor1$ will open the fingers. Write a PLC program to operate the gripper.

6. A quality inspection machine uses a laser to scan the surface of a product to identify defects. The machine is an XY table with two axes. The laser is positioned at a fixed height from the surface of the part. The scanner uses the pattern shown in Figure 7.25. Write a motion program similar to Algorithm 6 to scan the part. Each cycle of the machine starts and finishes at the home position H. The scan velocity needs to be 5 in/s.

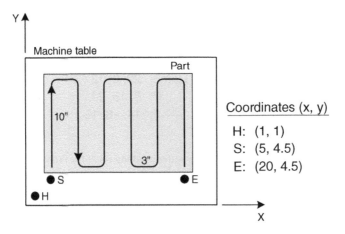

Figure 7.25   Laser scanner for surface defects for Problem 6

7. The automatic bottle filling machine in Figure 7.26 has two linear axes (X and Z) and a solenoid valve. The nozzle is inserted into the bottle until its tip is 0.5 in away from the bottom of the bottle. The solenoid valve is then turned on allowing the liquid to flow into the bottle. The goal is to maintain constant distance between the tip of the nozzle and the surface of the liquid in the bottle as it is filling up. Due to the tapered bottle, the nozzle must move up faster and faster to maintain a constant distance between the rising liquid surface and the tip of the nozzle. When the liquid reaches the bottle neck, the solenoid valve is shut off, which stops the flow immediately. The machine uses a pump with constant flow rate of 1.5 in^3/s.

Write a motion program similar to Algorithm 6 to control this machine. The bottle is initially at the home position ($X = 0, Z = 0$). After filling, the full bottle is moved back to the home position. The solenoid valve is connected to $OUT1$ output.

Hint: Define a small time increment. During this increment, find how much the liquid will rise given the constant flow rate. Command the Z-axis to go up as much as this distance while the bottle is filling. This is similar to "cutting" the liquid in the bottle into horizontal

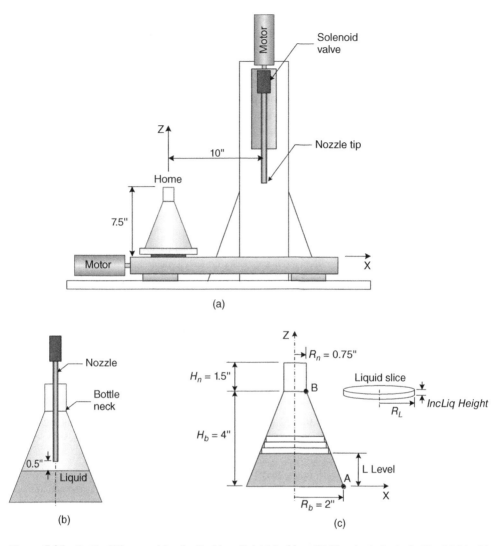

Figure 7.26   Bottle filling machine for Problem 7. (a) Machine. (b) Nozzle tip in the bottle. (c) Liquid "slice"

slices as shown in Figure 7.26c. Since the flow rate is constant, the slices get thicker as the liquid goes up. Each slice must be filled during the same time increment. Hence, the tip has to move faster and faster as the liquid rises. Consecutive motion commands to the Z-axis for each slice will be blended by the controller resulting in smooth motion of the axis.

8. A rotating knife similar to the one in Figure 7.13 is used to cut veneer wood products. The web moves at 60 in/s speed. The knife radius is 4 in. The cutting phase takes place over 4.2 in of web length ($L_c = 4.2$ in). Plot the knife rotational speed versus time if the desired cut length is (a) 18 in, and (b) 60 in. Each plot should be similar to Figure 7.14. Use 30 ms for acceleration time in blending and mark all pertinent data on the plots.

9. Write a motion control program for the rotating knife in Problem 8a with desired cut length of $L_d = 18$ in. Use the ratio following type of programming approach shown in Algorithm 7.

10. Aluminum foil is wound in a machine as shown in Figure 7.17. The unwind roll radius r_{roll} is measured by an ultrasonic sensor hooked up to an analog input of the motion controller. The nip rolls run at constant speed ω_{nip} and have a radius of r_{nip}. The rewind axis is operated in velocity mode with torque limit as given in Equation (7.11). To maintain relatively constant tension, the commanded velocity ω_{cmd} should be adjusted in real time based on the unwind roll radius. The roll diameter is measured by the ultrasonic sensor, which is an analog signal coming into the controller in analog input #1 ($AI1$). Write a motion control program to control the web tension.

References

[1] Application Note: SCARA Robot Kinematics. Technical Report, Delta Tau Data Systems Inc., 2004.

[2] Application Note: Spool Winding. Technical Report, Delta Tau Data Systems, Inc., 2004.

[3] Application Note #3415: Methods of Addressing Applications Involving Inverse Kinematics. Technical Report, Galil Motion Control, Inc., 2007.

[4] DMC-14x5/6 User Manual Rev. 2.7. Technical report, Galil Motion Control, Inc., 2011.

[5] Application Note, Flying Shear (AN00116-003). Technical Report, ABB Corp., 2012.

[6] Application Note, Rotary Axis Flying Shear (AN00122-003). Technical Report, ABB, 2012.

[7] ABB Corp. (2014) IRB 360 FlexPicker®. http://new.abb.com/products/robotics/industrial-robots/irb-360 (accessed 28 October 2014).

[8] Adept Technology, Inc. (2014) Adept Cobra i600 SCARA 4-Axis Robot. www.adept.com (accessed 3 August October 2014).

[9] Baldor. NextMove e100 Motion Controller.(2013) http://www.baldor.com/products/servo_drives.asp (accessed 14 August 2014).

[10] John Craig. *Introduction to Robotics Mechanics and Control.* Pearson Education, Inc., Third edition, 2005.

[11] Jeff Damour. The Mechanics of Tension Control. Technical Report, Converter Accessory Corp., USA, 2010.

[12] Delta Tau Data Systems, Inc. *Turbo PMAC User Manual*, September 2008.

[13] Galil Motion Control, Inc. (2013) http://www.galilmc.com, (accessed 26 August 2014).

[14] Parker, Inc. *6K Series Programmer's Guide*, 2003.

[15] John Rathkey. Multi-axis Synchronization. (2013) http://www.parkermotion.com/whitepages/Multi-axis.pdf (accessed 15 January 2013).

[16] Rockwell Automation (2013). RSLogix 5000, Kinematics Features. http://www.rockwellautomation.com, 2013.

[17] Rockwell Automation (2014). ControlLogix Control System. http://ab.rockwellautomation.com/Programmable -Controllers/ControlLogix#software (accessed 24 November 2014).

[18] Rockwell Automation (2014). RSLogix 5000, Design and Configuration. http://www.rockwellautomation.com /rockwellsoftware/design/rslogix5000/overview.page (accessed 24 November 2014).

[19] Siemens AG. SIMOTION Software (2014). http://w3.siemens.com/mcms/mc-systems/en/automation-systems /mc-system-simotion/motion-control-software/Pages/software-iec-61131.aspx (accessed 12 November 2014).

[20] Yaskawa America, Inc. 2013 Kinematix Toolbox, MP2300Siec Motion Controller. http://www.yaskawa.com (accessed 18 September 2014).

Appendix A

Overview of Control Theory

This appendix presents a very brief review of basic definitions and tools for the study of control systems. There is a vast amount of theory and many books on control systems such as [1, 3, 4].

A.1 System Configurations

Control systems are configured either as open-loop or closed-loop systems.

Open-loop systems do not measure the actual output of the system. The main advantage is the lower cost. However, open-loop systems cannot compensate for any *disturbances*, which are unexpected inputs to the system.

Closed-loop systems measure the actual system output and continuously compare it to the desired system output (Figure A.1). Based on the difference (*error*), they can adjust the control action needed to closely follow the desired output (*reference input*). These systems are also called *feedback control systems*. Measuring and feeding the actual output back to the system is called the *feedback* or the *loop*.

The main advantages of feedback systems are greater accuracy and less sensitivity to disturbances compared to the open-loop systems. Feedback systems are harder to design and are more expensive to implement than the open-loop systems.

A.2 Analysis Tools

Mathematical models are used in the analysis of control systems. Dynamic system behavior is governed by differential equations. Laplace transform ($\mathcal{L}[f(t)]$) enables us to convert these equations into algebraic equations. Furthermore, it allows separating system components or subsystems into individual entities with I/O relationships. Some basic Laplace transforms are given in Table A.1. These are used when taking the Laplace of differential equations. Table A.2 shows basic Laplace transforms of two time functions which correspond to two types of inputs frequently used in the analysis of control systems.

Industrial Motion Control: Motor Selection, Drives, Controller Tuning, Applications, First Edition. Hakan Gürocak.
© 2016 John Wiley & Sons, Ltd. Published 2016 by John Wiley & Sons, Ltd.

OPEN-LOOP SYSTEM

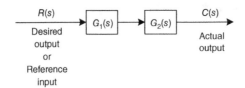

CLOSED-LOOP SYSTEM

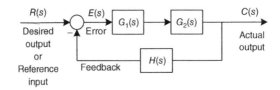

Figure A.1   Open-loop and closed-loop control systems

Table A.1   Laplace transform theorems

Theorem	Definition
$f(t)$	$F(s)$
$\mathcal{L}\left[\frac{df}{dt}\right] = sF(s)$	Differentiation[a]
$\mathcal{L}\left[\frac{d^2f}{dt^2}\right] = s^2 F(s)$	Differentiation[a]
$\mathcal{L}\left[\int_{0-}^{t} f(\tau)d\tau\right] = \frac{F(s)}{s}$	Integration
$f(\infty) = \lim\limits_{s \to 0} sF(s)$	Final value theorem

[a]With zero initial conditions.

Table A.2   Basic Laplace transforms

Time function	Laplace transform
$f(t)$	$F(s)$
Unit step, $u(t) = 1$	$\frac{1}{s}$
Ramp, $tu(t)$	$\frac{1}{s^2}$

Figure A.2   Transfer function represented as a block

A.2.1   Transfer Functions

A transfer function (TF) describes the relationship between the input and output of a component, subsystem or an entire system. It is defined as:

$$TF = \frac{\text{output}}{\text{input}} \tag{A.1}$$

A transfer function can be represented by a block as shown in Figure A.2. Here, the input signal is multiplied by the transfer function to convert it into the output of the block. In many textbooks, $G(s)$ is used as a generic representation for a transfer function like the ones shown in Figure A.1.

■ EXAMPLE A.2.1

The equation of motion for a spring-mass-damper system is given with $x(t)$ and $f(t)$ representing the position of the mass and the external force, respectively. Find the transfer function for this system if $x(t)$ is the output resulting from the $f(t)$ input.

$$m\frac{d^2x}{dt} + b\frac{dx}{dt} + kx = f(t)$$

Solution
Taking Laplace transform of both sides using the theorems in Table A.1 gives:

$$(ms^2 + bs + k)X(s) = F(s)$$

Then, the transfer function can be found using the definition in Equation (A.1) as:

$$TF = \frac{X(s)}{F(s)} = \frac{1}{ms^2 + bs + k} \tag{A.2}$$

A.2.1.1   Poles and Zeros

This terminology is used to refer to the roots of the numerator and denominator of a transfer function.

   Poles are the roots of the denominator of a transfer function. They describe the dynamic behavior of the system governed by the transfer function.

   Zeros are the roots of the numerator of a transfer function. They can affect the amplitude of response of a system.

■ EXAMPLE A.2.2

Given the following transfer function. Find the poles and zeros.

$$G(s) = \frac{s+3}{s^2 + 2s + 9} \tag{A.3}$$

Solution

First, we can equate the numerator polynomial to zero and find its roots (zeros). In this case, there is only one zero at $z = -3$. Note that not all systems will have zeros such as the one given in Equation (A.2).

Next, we can equate the denominator polynomial to zero and find its roots (poles). In this case, they turn out to be complex numbers as $s_1 = -1 + j\sqrt{8}$ and $s_2 = -1 - j\sqrt{8}$.

A.2.2   Block Diagrams

In control systems, block diagrams are used to describe the interconnections between components and subsystems of the entire system. Blocks can be interconnected using three basic connection types. Figure A.3 shows these connections and the corresponding simplified equivalent blocks.

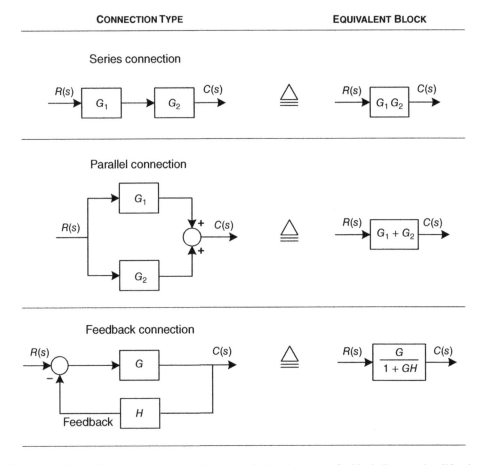

Figure A.3   Basic block connections and their equivalent replacements for block diagram simplification

A.2.2.1   Closed-Loop Transfer function (CLTF)

A feedback loop as shown in Figure A.3 can be replaced by an equivalent single block. The transfer function described by this block is called the *closed-loop transfer function (CLTF)*. It describes the dynamic behavior of the entire feedback loop. It is given by the following formula:

$$CLTF = \frac{G}{1 + GH} \tag{A.4}$$

■ **EXAMPLE A.2.3**

Find the CLTF for the closed-loop system shown in Figure A.1.

Solution

Since the $G_1(s)$ and $G_2(s)$ transfer functions are in series, they can be combined into a single block with $G = G_1 G_2$ in it. Then, with this new transfer function, the system looks just like the feedback connection in Figure A.3. Using the formula in (A.4), we can find

$$CLTF = \frac{C(s)}{R(s)} = \frac{G_1 G_2}{1 + G_1 G_2 H}$$

A.3   Transient Response

Step input represents a sudden change in the input to a control system. Often *unit* step input is used to analyze the system response. From Table A.2, the Laplace transform of a unit step input is equal to $\frac{1}{s}$.

Ramp input represents a constant change in the input to a control system. For example, velocity with constant acceleration in a motion control system is a ramp input. From Table A.2, the Laplace transform of a ramp input with *unit* slope is equal to $\frac{1}{s^2}$.

A system can be classified as first-order or second-order based on the power of the highest s term in the denominator of the transfer function.

A.3.1   *First-Order System Response*

The denominator of a first-order system has "s" as the highest power term. Figure A.4 shows the response of a generic first-order system given by:

$$G(s) = \frac{K}{s + a}$$

Time constant τ is used to describe the speed of the response. It is inversely proportional to the pole location "a" ($\tau = \frac{1}{a}$). As "a" gets larger, the time constant gets smaller and the system responds faster.

A.3.2   *Second-Order System Response*

The denominator of a second-order system has s^2 as the highest power term. A generic second-order system transfer function is:

$$G(s) = \frac{K}{as^2 + bs + c}$$

where $K, a, b,$ and c are all constants.

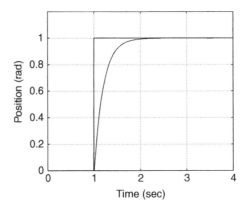

Figure A.4   Step response of a first-order system

Standard form for a second-order system is given by the following [3]:

$$G(s) = \frac{\omega_n^2}{s^2 + 2\zeta\omega_n s + \omega_n^2}$$

where ω_n is the *natural frequency* and ζ is the *damping ratio*. The damping ratio is an indication of how fast oscillations will die out in the response of a system. The natural frequency is the oscillation frequency of the system if there were no damping.

As shown in Figure A.5, four different types of response are possible with the second-order systems depending on the coefficients of the denominator polynomial. They can be categorized based on the damping ratio as

1. Undamped ($\zeta = 0$),
2. Underdamped ($0 < \zeta < 1$),
3. Critically damped ($\zeta = 1$), and
4. Overdamped ($\zeta > 1$).

■ **EXAMPLE A.3.1**

Find the unit step response of a system with unity feedback and $G(s) = \frac{50}{s(s+3)}$.

Solution
We need to first replace the feedback system with its equivalent CLTF. Using Equation (A.4), we can find the CLTF as

$$CLTF = \frac{C(s)}{R(s)} = \frac{50}{s^2 + 3s + 50} \tag{A.5}$$

To find the unit step response, we can insert $R(s) = \frac{1}{s}$. Then,

$$C(s) = \frac{50}{s^2 + 3s + 50} \cdot \frac{1}{s} \tag{A.6}$$

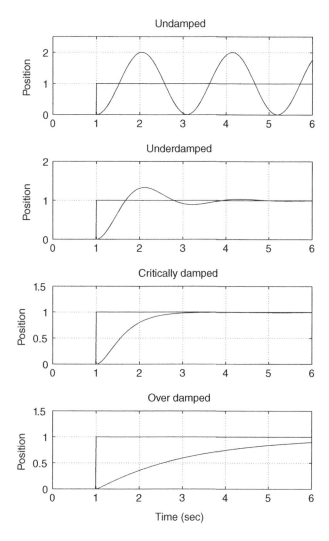

Figure A.5   Four different types of possible step responses of second-order systems

This equation can be solved using partial fraction and inverse Laplace transforms. The solution can also be obtained numerically using simulation. Figure A.6 shows a Simulink® model for the system and the resulting response.

A.3.2.1   Underdamped Response Specifications

Many physical systems have underdamped response. Several parameters were defined to describe the specifications of an underdamped response as shown in Figure A.7.

Peak time (T_p) is the time it takes to reach the first peak of the response. *Percent overshoot* (%OS) is how far the system goes past the desired final value at its first peak. It is calculated

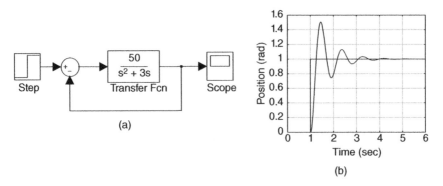

(a)

(b)

Figure A.6 Transfer function and step response of the second-order system in Example A.3.1. (a) Simulation model for the system. (b) Unit step response

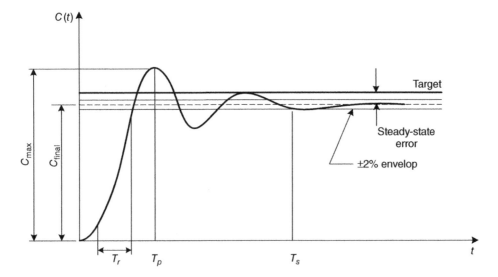

Figure A.7 Specifications for underdamped response

as a percentage as follows:

$$\%OS = \frac{C_{max} - C_{final}}{C_{final}} \times 100$$

Settling time (T_s) is the time it takes for the response to reach *and stay* within a $\pm 2\%$ envelop at the steady-state value. *Rise time* (T_r) is the time required for the response to go from 10% to 90% of the final value. The rise time and settling time provide information about the speed of the response. Percent overshoot is an indication of stability of the system.

Bandwidth is a frequently used term in control systems. It is measured in hertz (Hz). It indicates the capability of a system to follow rapidly changing input. Bandwidth is a range of frequencies. Any input changing with a frequency within the bandwidth can be successfully followed by the system without falling behind or limiting its output amplitude. The following is an approximate relationship between the bandwidth and the settling time [2]:

$$f_{BW} \approx \frac{4}{2\pi T_p}$$

where f_{BW} is the bandwidth in Hz. For example, if a system has 0.01 s settling time, then its bandwidth is 63.7 Hz. In other words, it can successfully follow any input that is changing with a frequency of up to 63.7 Hz.

A.4 Steady-State Errors

Steady-state is where the system settles after its transient response to an input such as a step input. Depending on the combination of the input and system type, there may be a difference (error) between where the system actually settles versus the desired target (Figure A.7). For example, in a motion control application, if an axis was commanded to move 10 in and it stops after moving 9.8 in, there will be a steady-state error of 0.2 in.

In studying steady-state errors, systems can be classified into *Type 0*, *Type 1*, or *Type 2* based on the number of integrators in the forward path (going from input $R(s)$ to output $C(s)$ in Figure A.3). In the Laplace domain, an *integrator* is equal to $\frac{1}{s}$ as shown by the integration entry in Table A.1. A generic system transfer function is given by:

$$G(s) = \frac{K(s + z_1)(s + z_2) \dots}{s^n(s + p_1)(s + p_2) \dots} \tag{A.7}$$

where K is a constant called *gain*, $z_1, z_2 \dots$ are the zeros and $p_1, p_2 \dots$ are the poles of the system. The s^n term represents n integrators in the forward path. Therefore, for Type 0 systems $n = 0$. Type 1 and 2 systems have $n = 1$ and $n = 2$, respectively.

Often control systems are configured to have unity feedback ($H(s) = 1$ in Figure A.3). Table A.3 shows the relationship between the system type, input, and the resulting steady-state error. For example, from Table A.3, a Type 1 system will have no steady-state error if it is given a step input. But the same system will have a constant following error if it is given a ramp input. The entries in this Table are found from analyzing the error transfer function of a system with different input types using the *final value theorem* in Table A.1.

Table A.3 Steady-state errors as a function of input and system type combinations

Input	System type		
	Type 0	Type 1	Type 2
Step	C	0	0
Ramp	∞	C^a	0

^{a}C: Constant

References

[1] Richard Dorf and Robert Bishop. *Modern Control Systems*. Prentice-Hall, Twelfth edition, 2011.

[2] George Ellis. *Control Systems Design Guide*. Elsevier Academic Press, Third edition, 2004.

[3] Norman Nise. *Control Systems Engineering*. John Wiley & Sons, Inc., 2007.

[4] Katsuhiko Ogata. *Modern Control Engineering*. Prentice-Hall, Fifth edition, 2009.

Index

Industrial Motion Control: Motor Selection, Drives, Controller Tuning, Applications, First Edition. Hakan Gürocak.
© 2016 John Wiley & Sons, Ltd. Published 2016 by John Wiley & Sons, Ltd.